AF777406

Principles of Differential and Integral Equations

Principles of Differential and Integral Equations

C. Corduneanu

University of Jassy
Jassy, Romania

Chelsea Publishing Company
The Bronx, New York

Second Edition

© Copyright, 1971, by Constantin Corduneanu

© Copyright, 1977, by Constantin Corduneanu

The second edition differs from the first in various revisions throughout the text and bibliography. It is published at The Bronx, New York and is printed on the recently developed 'long-life' acid-free paper.

Library of Congress Cataloging in Publication Data

Principles of differential and integral equations.

Bibliography: p.
Includes index.
1. Differential equations. 2. Integral Equations. I. Title.
QA371.C77 1977 515′.35 77-2962
ISBN 0-8284-0295-7

Printed in the United States of America

To the Memory of

My Mother

Preface

This book is intended to provide sufficient material for a three-hour one-semester course in the theory of differential and integral equations, and is addressed primarily to undergraduate students at the intermediate level.

Usually, any elementary course provides enough material related to the algorithmic aspects of theory (how to solve an equation). In such a course, many classes of equations are solved by means of formulas, using series expansions, Laplace transform, etc. Also, one finds various motivations and illustrations from physics, engineering science, and related fields. Yet, only rarely does the reader find in an elementary course some comments concerning the general problem of existence, the uniqueness problem, the behavior of solutions, and several other properties that are interesting in themselves or with respect to their applications.

The aim of this book is to offer a discussion of the fundamental concepts in the theory of differential and integral equations, with a view toward preparing the reader for graduate-level courses on advanced topics. This book can also serve a graduate-level course for non-mathematics majors. Students of various disciplines have found an increasing need to study such topics as control theory and systems theory, which are mainly based on differential equations, integral operators, and related concepts. The material included in this text will provide sufficient basis for understanding these new areas of investigation.

The prerequisites are a good course in advanced calculus (set theory and elementary topological concepts, properties of continuous functions, uniform convergence, implicit functions), some preparation in linear algebra, and a reasonable acquaintance with elementary complex analysis (including entire and meromorphic functions). No prerequisite is needed in differential or integral equations! Indeed, the book can be used as a first course in differential and integral equations by any reader possessing knowledge of the above-mentioned fields.

To aid the reader, the text includes several auxiliary results from algebra and calculus, some with complete proofs and the others with adequate references. Each chapter ends with a number of problems. Very often these problems are quite challenging, representing extensions of the theory beyond the level reached in the text rather than standard applications of some formulas or theorems.

Despite the fact that by their inner structure the two theories to which this book is dedicated are quite different, hopefully the reader will not be shocked in passing from one subject to the other. Thus, the theory of Volterra integral equations appears as a direct continuation of the theory of ordinary differential equations. Moreover, some topics used in the theory of differential equations apply to Fredholm integral equations, and the most salient results in the theory of integral equations with symmetric kernel lead to an elegant solution of boundary value problems for ordinary differential equations (Sturm-Liouville problems).

Finally, use of the Lebesgue integral was avoided—not because this tool would be useless in studying differential or integral equations but because it is unwieldy for many students.

The author expresses appreciation to his colleagues and students who have contributed to the preparation of this book: Viorel Barbu, Ioana Iancu, Nicolae Pavel, and Pavel Talpalaru, all with the Department of Mathematics (Seminarul Matematic) of the University of Jassy. Thanks go also to Mr. Teoctist Galan and Miss Roza Oxenberg for their unfailing cooperation in preparing the manuscript. Finally, it is with pleasure that the author expresses sincere thanks to Professor Dan Petrovanu for his invaluable assistance in checking the proof.

C. Corduneanu
Jassy, May 1969

Contents

Introduction

Differential equations as well as integral equations belong to a wider class of equations in which the unknown object is a function (scalar function or vector function). Such kinds of equations are often encountered in mathematics and in various sciences that use the mathematical apparatus, and they are generally called *functional equations.*

Among functional equations, the *differential* ones can be recognized by the fact that some derivatives of the unknown function appear in the equation. For instance,

$$F(t, x(t), \dot{x}(t)) = 0 \tag{1}$$

represents a differential equation in which $F(t, x, y)$ is a given function, $x(t)$ is the unknown function of the independent variable t, and $\dot{x}(t)$ denotes the first derivative of the function $x(t)$.

An *integral equation* may be described as being a functional equation involving the unknown function under one or more integrals. For example,

$$F\left(t, x(t), \int_a^b k(t, s)x(s)\, ds\right) = 0 \tag{2}$$

where $k(t, s)$ and $F(t, x, y)$ are given functions and $x(t)$ is the unknown, offers an illustration of the concept of integral equation.

Certainly, the reader is aware of a class of functional equations that appear in any elementary textbook on calculus. We have in mind the equation defining the so-called implicit functions. Such an equation has the form

$$f(t, x(t)) = 0 \tag{3}$$

and the following result gives an existence theorem for (3).

Assume that $f(t, x)$ and $\partial f(t, x)/\partial x$ are continuous functions in a neighborhood of (t_0, x_0). Moreover, assume $f(t_0, x_0) = 0$ and $\partial f(t, x)/\partial x \neq 0$ at (t_0, x_0). Then, there exists one and only one continuous function $x(t)$, defined for $|t - t_0|$ sufficiently small, satisfying (3) *and such that $x(t_0) = x_0$.*

Many results concerning the existence of solutions of various classes of functional equations can be derived from a more general form of the implicit function theorem. We are not going to develop here this approach to the existence theory of functional equations. It requires some preparation in functional analysis, and, on the other hand, the direct investigation of various classes of functional equations gives the possibility of finding better results.

The basic problems arising in connection with a given functional equation are the following:

1. The *existence* of solutions;
2. The *uniqueness* problem—that is, what conditions should be added in order to determine a single solution;
3. The *construction* of the solution.

Each one of these problems can be understood in a variety of ways. For instance, the differential equation $\dot{x} = 1 + x^2$ has no real solutions defined on an interval $[t_0, t_0 + \pi]$, with t_0 arbitrarily chosen in the real field. It has infinitely many solutions defined on $[t_0 + \varepsilon, t_0 + \pi]$, where $\varepsilon > 0$ is a small number. Indeed, $x(t) = \tan(t + C)$ with arbitrary real C represents all the solutions of the given equation, and the preceding statement can be easily verified. Concerning the third problem, we notice that the word "construction" needs to be made more precise. For example, we could mean a representation of the solutions by means of a formula involving the data of the problem (see Chap. 1). Also, we could think to express the solution as the sum of a series or the limit of a convergent sequence of functions, and so on. Very often, a numerical approach is quite satisfactory (i.e., to find approximately the values of the solution for some prescribed values of the independent variable).

We shall consider many other interesting problems in this book. For instance, once the existence and uniqueness of the solution are proved, an important problem is to find the maximum interval on which this solution can be extended. It is also interesting to know whether this unique solution changes continuously with respect to the variation of some data involved in the equation, and so on. When we are dealing with solutions defined on a half-axis or on the whole real axis, an important problem is to get some information concerning the behavior "at infinity" (e.g., the boundedness of the solution, the exponential decay, and other kinds of behavior).

Several types of equations (differential or integral) depend on parameters. The discussion of these equations with respect to the parameters offers another kind of problem often encountered in the theory and its applications.

Chapter 1 of this book is devoted to the discussion of some elementary types of differential equations and related topics. Inasmuch as we can illustrate the uniqueness problem by using a very special kind of differential equation with separable variables, we pay much attention to this problem. We shall consider it in the sequel, in view of further development. Elementary differential and integral inequalities are also discussed, in view of their application to some basic problems.

The second chapter contains the existence theory for ordinary differential equations and systems, as well as a uniqueness theorem of the Osgood type. The reader will find in this chapter a local theory of existence, the main results being related to the method of successive approximations (usually called Picard's method) and to the Ascoli–Arzelà criterion of compactness of families of continuous functions. For the reader's convenience, we give the proof of this criterion, which is used frequently in this book.

Some global problems in the theory of ordinary differential equations are considered in the third chapter. Such problems as global uniqueness, continuation of solutions, behavior of a saturated solution at the endpoints of its interval of existence, and dependence of solutions on their initial values are discussed. This chapter contains also an introduction to the comparison technique in the theory of ordinary differential equations and its application to the global existence problem.

Chapter 4 is concerned with linear systems and equations—except one section, which contains a discussion of the autonomous systems.

The stability theory of differential systems is sketched in the fifth chapter. The reader will find in this chapter the basic concepts and results pertaining to the general theory of stability, as well as a relevant application in the study of automatic control systems. Several problems at the end of this chapter will stimulate the reader in view of further developments in this important branch of modern research.

Chapter 6 is dedicated to the theory of Volterra integral equations. The last section of this chapter gives an idea of the possibility of using the theory of Volterra integral equations to find results concerning the differential systems (mainly global aspects).

In the seventh chapter can be found a general discussion of Fredholm integral equations and some auxiliary facts related to the Hilbert space theory.

The eighth chapter is devoted to the theory of self-adjoint integral equations. This theory gives us the opportunity to solve the important problem of existence of eigenvalues and eigenfunctions of Sturm–Liouville systems, which appears, for instance, in quantum mechanics [let us recall the famous paper by E. Schrödinger—"Quantisierung als Eigenwertproblem," *Ann. Physik* 79 (1926)].

Finally, the last chapter contains several selected topics which, we hope, will stimulate the reader's interest in further study of related problems.

1

Some Elementary Methods

Several kinds of differential equations can be solved by elementary methods. More precisely, for such types of equations as linear differential equations of the first order, equations with separable variables, and equations that can be reduced to them by appropriate transformations, it is possible to express the solution by a formula involving the data of the problem. Under convenient conditions, we can prove the existence and uniqueness of the solution, as well as some other useful properties.

The main significance of these elementary methods consists primarily in the fact that the solution can be explicitly expressed. However, it does not necessarily mean that in such a case we are facing a very simple problem. The formulas occurring in the application of these methods may require, beside the integration of some functions, such operations as finding the inverse of a given function, which cannot be carried out by "elementary" methods.

§1.1. The Linear Equation of the First Order

The general linear equation of the first order has the form

$$\dot{x} = a(t)x + b(t), \qquad t \in (t_1, t_2) \tag{1.1}$$

where $a(t)$ and $b(t)$ are assumed continuous on (t_1, t_2). Let us remark that the cases $t_1 = -\infty$ or $t_2 = +\infty$ are not excluded from our considerations. In what follows, we shall assume that $t \in (t_1, t_2)$. We associate to Eq. (1.1) the "initial condition,"

$$x(t_0) = x_0, \qquad x_0 \in R. \tag{1.2}$$

To construct the solution (or solutions) of our problem, we shall assume first its existence on an interval containing t_0. Equation (1.1) can be written in the form

$$\dot{x} - a(t)x = b(t)$$

and by multiplication with $\exp[-\int_{t_0}^t a(s)\,ds]$ it becomes

$$\frac{d}{dt}\left\{x(t)\exp\left[-\int_{t_0}^t a(s)\,ds\right]\right\} = \exp\left[-\int_{t_0}^t a(s)\,ds\right]b(t).$$

Integrating from t_0 to t, with t belonging to the interval of definition of $x(t)$, we get

$$x(t)\exp\left[-\int_{t_0}^t a(s)\,ds\right] - x(t_0) = \int_{t_0}^t \exp\left[-\int_{t_0}^u a(s)\,ds\right]b(u)\,du.$$

Taking (1.2) into account, we have

$$x(t) = \exp\left[\int_{t_0}^t a(s)\,ds\right]\left\{x_0 + \int_{t_0}^t \exp\left[-\int_{t_0}^u a(s)\,ds\right]b(u)\,du\right\} \tag{1.3}$$

or, in an equivalent form,

$$x(t) = x_0 \exp\left[\int_{t_0}^t a(s)\,ds\right] + \int_{t_0}^t \exp\left[\int_u^t a(s)\,ds\right]b(u)\,du. \tag{1.4}$$

Formula (1.3) [or (1.4)] shows that $x(t)$ is uniquely determined by (1.1) and (1.2)—that is, we proved the uniqueness of the solution and found it, under the assumption of its existence.

Consider now the function $x(t)$ given by (1.3), and show that it represents a solution of Eq. (1.1), with the initial condition (1.2), defined on the whole interval (t_1, t_2). Indeed, from (1.3) it follows that $x(t)$ is defined on (t_1, t_2), because $\int_{t_0}^t a(s)\,ds$ and the integral in the braces are both defined on (t_1, t_2). It is obvious that $x(t_0) = x_0$—that is, condition (1.2) is satisfied. It remains to prove that $x(t)$ satisfies (1.1). First, we remark that $x(t)$ is a differentiable function on (t_1, t_2), both factors in (1.3) being differentiable. A simple calculation gives

$$\dot{x}(t) = a(t)\exp\left[\int_{t_0}^t a(s)\,ds\right]\left\{x_0 + \int_{t_0}^t \exp\left[\int_u^t a(s)\,ds\right]b(u)\,du\right\}$$
$$+ \exp\left[\int_{t_0}^t a(s)\,ds\right]\cdot\exp\left[-\int_{t_0}^t a(s)\,ds\right]b(t)$$

which can be written as

$$\dot{x}(t) = a(t)x(t) + b(t).$$

That is, we obtained (1.1).

Consequently, we can formulate the following result.

Theorem 1.1

Consider Eq. (1.1) *with continuous* $a(t)$ *and* $b(t)$ *on* (t_1, t_2). *Then, there exists a unique solution of this equation, defined on* (t_1, t_2) *and satisfying* (1.2). *It is given by the formula* (1.3) *or* (1.4).

It is worth pointing out the following feature of the preceding existence result: The solution has the same interval of definition as the coefficients $a(t)$ and $b(t)$. This property does not remain valid for nonlinear equations (in general). For instance, the equation $\dot{x} = -x^2$ has the solutions $x(t) = (t - t_0)^{-1}$, $t_0 \in R$, and $t \neq t_0$, and none of them is defined on the whole real axis (as the right-hand side of the equation).

Remark. Consider the differential inequality associated with (1.1), namely

$$\dot{y}(t) \leq a(t)y(t) + b(t), \qquad t \in [t_0, t_2). \tag{1.5}$$

If we know that $y(t)$ satisfies also

$$y(t_0) \leq x_0 \tag{1.6}$$

then we get

$$y(t) \leq x(t), \qquad t \in [t_0, t_2) \tag{1.7}$$

where $x(t)$ is given by (1.3).

To prove (1.7), we can follow the same method as previously {i.e., multiply (1.5) by the *integrating factor* $\exp[-\int_{t_0}^{t} a(s)\,ds]$ and integrate from t_0 to $t \in (t_0, t_2)$}.

We would like to mention that (1.7) should be understood as follows: Any differentiable function $y(t)$ that satisfies (1.5) and (1.6) also satisfies (1.7). Of course, not every function satisfying (1.7) is a solution of (1.5). In other words, nothing is said about the set of solutions of the inequality (1.5), more than (1.7) (e.g., how rich it is).

§1.2. The Equation with Separable Variables

Another type of first-order differential equation, frequently encountered in the applications, is that of *equation with separable variables*; that is,

$$\dot{x} = f(t)g(x). \tag{1.8}$$

We shall study Eq. (1.8) under the following assumptions:

(a) $f(t)$ is a continuous function on the interval (t_1, t_2), the possibilities $t_1 = -\infty$ or $t_2 = +\infty$ being accepted;

(b) $g(x)$ is a continuous function on the interval (x_1, x_2), with $x_1 = -\infty$ or $x_2 = +\infty$ allowed, such that $g(x) \neq 0$ on (x_1, x_2).

Let us remark that $g(x)$ should preserve the sign on (x_1, x_2), because it is assumed continuous. Without loss of generality, we can suppose $g(x) > 0$. Otherwise, we have to change the signs of both factors $f(t)$ and $g(x)$ in (1.8), which leaves the equation unchanged.

The initial condition (1.2) would be associated now to Eq. (1.8), with

$$t_0 \in (t_1, t_2), \qquad x_0 \in (x_1, x_2). \tag{1.9}$$

The main result concerning Eq. (1.8) can be stated as follows.

Theorem 1.2

Consider Eq. (1.8), *under assumptions* (*a*) *and* (*b*) *stated previously. There then exists a unique solution* $x(t)$ *of this equation, satisfying also* (1.2), *and defined in a neighborhood of* t_0. *This solution is given by*

$$x(t) = G^{-1}\left(G(x_0) + \int_{t_0}^{t} f(s)\, ds\right) \tag{1.10}$$

where $G(u)$ *is a primitive of the function* $1/g(u)$ *on* (x_1, x_2) *and* G^{-1} *denotes the inverse of* G.

Proof. Assume there is a function $x(t)$ defined in a neighborhood of t_0 and satisfying (1.8):

$$\dot{x}(t) = f(t)g(x(t)).$$

Inasmuch as $g(x) \neq 0$, we can write the preceding equation in the equivalent form

$$\frac{\dot{x}(t)}{g(x(t))} = f(t)$$

or

$$\frac{dx(t)}{g(x(t))} = f(t)\, dt.$$

Integrating now from t_0 to t, with t belonging to the neighborhood of t_0 where the solution is defined, we get

$$\int_{x_0}^{x(t)} \frac{du}{g(u)} = \int_{t_0}^{t} f(s)\, ds \tag{1.11}$$

taking into account condition (1.2) and making the change of variable $u = x(t)$ in the first integral.

If $G(u)$ is such that

$$G'(u) = \frac{1}{g(u)}, \qquad u \in (x_1, x_2) \tag{1.12}$$

then (1.10) is equivalent to

$$G(x(t)) = G(x_0) + \int_{t_0}^{t} f(s)\, ds. \tag{1.13}$$

On the other hand, $G(u)$ is strictly increasing on (x_1, x_2), which implies the existence of a unique inverse G^{-1}. Consequently, (1.13) is equivalent to (1.10).

Because (1.10) contains G, which is not uniquely determined by (1.12), we cannot conclude that the solution is unique. But this difficulty is only apparent, because (1.13) is equivalent to

$$H(x(t)) = H(x_0) + \int_{t_0}^{t} f(s)\, ds \tag{1.14}$$

where $H(u)$ is another primitive function for $1/g(u)$. It yields

$$x(t) = H^{-1}\left(H(x_0) + \int_{t_0}^{t} f(s)\, ds\right) \tag{1.15}$$

which is equivalent to (1.14), hence to (1.10). In other words, $x(t)$ is the same, no matter how we choose the primitive function for $1/g(u)$, which proves the uniqueness. Consequently, if we assume the existence of a solution of (1.8), with initial value (1.2), we find that it is uniquely determined and is expressed by (1.10), whereas $G(u)$ is as previously described.

We consider now the function $x(t)$ given by (1.10), or equivalently by (1.13), and let us prove that it is defined in a neighborhood of t_0 and satisfies both (1.8) and (1.2).

Indeed, G^{-1} has as its domain of definition the range of the function $G(u)$, $u \in (x_1, x_2)$. Inasmuch as $G(u)$ is strictly increasing, it follows that its range is an open interval. It means that $G(x_0)$ belongs to the range of $G(u)$, together with a sufficiently small neighborhood, say V. We shall have $G(x_0) + \int_{t_0}^{t} f(s)\, ds \in V$ for $|t - t_0|$ sufficiently small, that is, for t belonging to a neighborhood of t_0. More precisely, $x(t)$ is defined for those values of t such that

$$\lim_{u \to x_1+} G(u) < G(x_0) + \int_{t_0}^{t} f(s)\, ds < \lim_{u \to x_2-} G(u).$$

Obviously, (1.2) is satisfied.

To prove that (1.8) is also satisfied, we shall use (1.13) instead of (1.10). It follows that

$$G'(x(t))\dot{x}(t) = f(t)$$

and, taking (1.12) into account, we get

$$\dot{x}(t) = f(t)g(x(t))$$

which proves the theorem.

Remark. Formula (1.10) involves two operations of integration, namely, the integration of the functions $f(t)$ and $1/g(x)$ and the calculation of the inverse function $G^{-1}(u)$. Consequently, the integration of an equation with separable variables requires more operations than the integration of a linear first-order equation.

§1.3. Exact Differential Equations

It is our aim to study the differential equation

$$M(t, x)\,dt + N(t, x)\,dx = 0 \tag{1.16}$$

under the assumption that its left-hand side is the exact differential of a function $F(t, x)$:

$$dF(t, x) = M(t, x)\,dt + N(t, x)\,dx. \tag{1.17}$$

If there exists a function $F(t, x)$ such that (1.17) holds in a given domain, then Eq. (1.16) is called an *exact differential equation.*

Let us formulate now the conditions under which Eq. (1.16) is an exact differential equation and use this fact to integrate it.

Theorem 1.3

Consider Eq. (1.16) *with* $M(t, x)$, $N(t, x)$, $\partial M/\partial x$, *and* $\partial N/\partial t$ *continuous in a neighborhood of the point* (t_0, x_0).

If

$$\frac{\partial M(t, x)}{\partial x} = \frac{\partial N(t, x)}{\partial t} \tag{1.18}$$

in the given neighborhood and $N(t_0, x_0) \neq 0$, *then there exists a unique solution* $x(t)$ *of Eq.* (1.16) *satisfying the initial condition* $x(t_0) = x_0$ *and defined in a neighborhood of* t_0.

Proof. Under assumption (1.18), we can take

$$F(t, x) = \int_{t_0}^{t} M(\tau, x)\,d\tau + \int_{x_0}^{x} N(t_0, \xi)\,d\xi \tag{1.19}$$

with (t, x) belonging to the neighborhood of (t_0, x_0), where $M(t, x)$ and $N(t, x)$ are given.

Now, it is obvious that (1.16) is equivalent to $F(t, x) = C$, with C an arbitrary constant. To fulfill the initial condition for $x(t)$, defined implicitly by $F(t, x) = C$, we have to take $C = F(t_0, x_0)$. Consequently, the function $x = x(t)$, defined by

$$F(t, x) - F(t_0, x_0) = 0 \tag{1.20}$$

is the solution of our problem.

That (1.20) indeed defines a function $x = x(t)$ can be derived easily from the existence theorem of implicit functions. We have $\partial F/\partial x = N \neq 0$ in a small neighborhood of (t_0, x_0), which is the key condition for the application of the previously mentioned theorem. The function $x(t)$, uniquely determined by (1.20), will be a differentiable function in a neighborhood of t_0. It satisfies (1.16) because it satisfies (1.20), an equivalent of (1.16). It satisfies also the required initial condition. Theorem 1.3 is completely proved.

Remark. Condition (1.18) guarantees that $F(t, x)$ given by (1.19) is such that (1.17) holds. It is satisfied only for a very "poor" class of differential equations of the form (1.16).

On the other hand, Eq. (1.16) is equivalent to any equation of the form

$$\rho(t, x)[M(t, x)\, dt + N(t, x)\, dx] = 0 \tag{1.21}$$

with $\rho(t, x) \neq 0$. Equation (1.21) is an exact equation if

$$\frac{\partial(\rho M)}{\partial x} = \frac{\partial(\rho N)}{\partial t}$$

or, in equivalent form,

$$N\frac{\partial \rho}{\partial t} - M\frac{\partial \rho}{\partial x} + \left(\frac{\partial N}{\partial t} - \frac{\partial M}{\partial x}\right)\rho = 0. \tag{1.22}$$

Consequently, if we can get a solution $\rho(t, x)$ of (1.22), such that $\rho(t, x) \neq 0$ in a certain domain, then Eq. (1.16) is reduced to an exact equation.

A function $\rho(t, x)$, with the preceding properties, is called an *integrating factor.*

For instance, in the case of the linear equation

$$dx - [a(t)x + b(t)]\, dt = 0$$

we used the integrating factor $\exp[-\int_{t_0}^{t} a(s)\, ds]$. In the case of an equation with separable variables,

$$dx - f(t)g(x)\, dt = 0,$$

the integrating factor was $1/g(x)$.

§1.4. The Uniqueness Problem: An Example

We have seen that the solution is unique if, besides our differential equation, we require the fulfillment of an initial condition. In the next chapters, we shall prove very general existence and uniqueness theorems. For instance, we shall establish the Peano's existence theorem for the differential equation $\dot{x} = f(t, x)$, with the initial condition $x(t_0) = x_0$, where $f(t, x)$ is merely continuous in a neighborhood of (t_0, x_0).

The uniqueness of an initial value problem, for the first-order differential equation $\dot{x} = f(t, x)$, requires something more than continuity for $f(t, x)$. We shall prove this assertion by considering the following simple equation:

$$\dot{x} = 3x^{2/3}. \tag{1.23}$$

The right-hand side of (1.23) is a continuous function for all real x, and because it is independent of t we can say that $f(t, x) = 2x^{2/3}$ is a continuous function on the whole plane (t, x).

It is obvious that (1.23) is an equation with separable variables. Nevertheless, the application of the theory developed in §1.2 is not possible without some precautions. Namely, the function $g(x) = 3x^{2/3}$ vanishes for $x = 0$. Consequently, we have to consider Eq. (1.23) separately, for $x > 0$ and for $x < 0$. It is interesting to remark that $x = 0$ is a solution of (1.23), defined for all real t.

Let us consider first the case $x > 0$. If we refer to our previous notations (see §1.2), we have $t_1 = -\infty$, $t_2 = +\infty$, $f(t) \equiv 1$, $x_1 = 0$, $x_2 = +\infty$, $g(x) = 3x^{2/3}$. We get $G(u) = u^{1/3}$, $G^{-1}(u) = u^3$, which yields

$$x(t) = (x_0^{1/3} + t - t_0)^3 \tag{1.24}$$

for any $x_0 > 0$, with arbitrary t_0.

A similar argument shows that (1.24) holds also for $x_0 < 0$. Moreover, any function $x(t)$ given by (1.24) is defined on the whole real axis. Inasmuch as $x_0^{1/3} - t_0$ is an arbitrary constant for x_0 and t_0 arbitrary, it follows that any function

$$x(t) = (t - \alpha)^3, \qquad \alpha = \text{real} \tag{1.25}$$

is a solution for (1.23), defined on the whole real axis.

Now, taking into account the fact that $x = 0$ is also a solution of (1.23), we can construct more sophisticated solutions. Namely, if α and β

are real numbers such that $\alpha \le \beta$, then

$$x(t) = \begin{cases} (t-\alpha)^3 & t < \alpha \\ 0 & \alpha \le t \le \beta \\ (t-\beta)^3 & t > \beta \end{cases} \tag{1.26}$$

is also a solution of (1.23), defined on the whole real axis. The graph of this function can be drawn easily, and is shown in Fig. 1. If $\alpha = \beta$, then (1.26) reduces to (1.25) and the graph is the cubic parabola defined by that equation.

That (1.26) defines a solution of (1.23) can be proved easily by starting from the preceding considerations. The only values of t that need some special considerations are $t = \alpha$ and $t = \beta$. At these points, the derivative could not exist. Fortunately, a simple inspection of $x(t)$ at $t = \alpha$ shows that $x(t)$ is differentiable, and its derivative vanishes at this point. The same is true for $t = \beta$.

Now we can answer the following question: How many solutions of (1.23) pass through a given point (t_0, x_0)?

Assume, for instance, that $x_0 > 0$. Then any solution of the form (1.26), with $\beta = t_0 - x_0^{1/3}$ and $\alpha \le \beta$, passes through (t_0, x_0). It is obvious that there exists an infinity of solutions passing through (t_0, x_0). The case $x_0 < 0$ leads to a similar conclusion.

Finally, in the case $x_0 = 0$, the situation is more typical for the non-uniqueness phenomenon. If we are interested in obtaining the solution of (1.23) such that $x(t_0) = 0$, for a given t_0 real, then we see that any solution

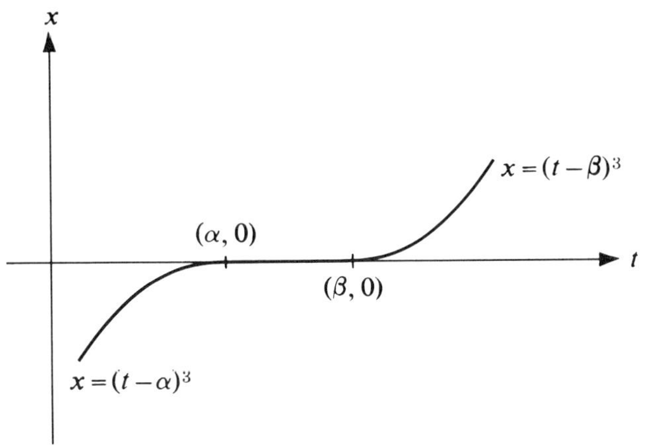

Figure 1

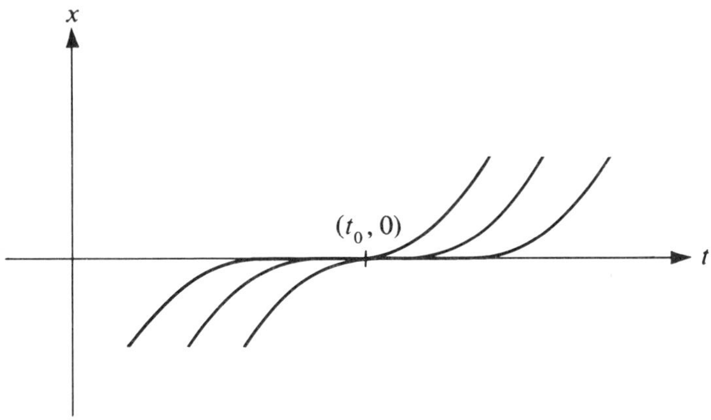

Figure 2

of the form (1.26), corresponding to $\alpha \leq t_0$ and $t_0 \leq \beta$, is convenient for us. The set of integral curves passing through $(t_0, 0)$ forms a *funnel*, as shown in Fig. 2.

If we want to go further with our discussion, it is interesting to remark that there exists a *maximal solution* passing through $(t_0, 0)$—namely,

$$x_M(t) = \begin{cases} 0 & t \leq t_0 \\ (t - t_0)^3 & t > t_0 \end{cases}$$

in the sense that the graph of any other solution passing through $(t_0, 0)$ lies under the graph of $x_M(t)$. In a similar manner, we can see that there exists a *minimal solution* $x_m(t)$.

§1.5. Some Integral Inequalities

Several problems we have to solve in the next chapters are particularly suitable for treatment based on the use of integral inequalities. We shall consider here two types of such inequalities, their solution being a simple consequence of the results established in the previous sections.

Consider first the linear inequality

$$x(t) \leq h(t) + \int_{t_0}^{t} k(s)x(s)\, ds, \qquad t \in [t_0, T) \tag{1.27}$$

where all the functions involved are continuous on $[t_0, T)$, $T \leq +\infty$, and $k(t) \geq 0$. The main result related to (1.27) can be stated as follows.

Lemma (Gronwall)

If the preceding assumptions hold for (1.27), *then* $x(t)$ *satisfies*

$$x(t) \le h(t) + \int_{t_0}^{t} h(s)k(s)\exp\left[\int_s^t k(u)\,du\right] ds, \qquad t \in [t_0, T). \quad (1.28)$$

Proof. Let us denote $y(t) = \int_{t_0}^t k(s)x(s)\,ds$ and remark that

$$\dot{y}(t) \le k(t)x(t) \le k(t)h(t) + k(t)y(t)$$

or

$$\dot{y}(t) \le k(t)y(t) + k(t)h(t). \quad (1.29)$$

Inequality (1.28) is of the form (1.5). Inasmuch as $y(t_0) = 0$, it follows [see (1.6) and (1.7)] that

$$y(t) \le \int_{t_0}^{t} h(s)k(s)\exp\left[\int_s^t k(u)\,du\right] ds, \qquad t \in [t_0, T)$$

which yields immediately (1.28).

A noteworthy particular case corresponds to $h(t) = M = \text{const}$—that is, inequality (1.27) becomes

$$x(t) \le M + \int_{t_0}^{t} k(s)x(s)\,ds, \qquad t \in [t_0, T). \quad (1.30)$$

In this case, we can integrate by parts in the right-hand side of (1.28) and obtain, after some obvious transformations,

$$x(t) \le M\exp\left[\int_{t_0}^{t} k(s)\,ds\right], \qquad t \in [t_0, T). \quad (1.31)$$

Another integral inequality we shall consider can be written in the form

$$x(t) \le M + \int_{t_0}^{t} k(s)g(x(s))\,ds, \qquad t \in [t_0, T). \quad (1.32)$$

This is, in general, a nonlinear inequality that reduces to (1.30) for $g(u) \equiv u$. As before, M is a constant, $k(t)$ is a continuous nonnegative function on $[t_0, T)$, and $g(x)$ is a continuous nondecreasing function such that $g(x) > 0$ on some half-axis $x \ge x_1$. The result we want to prove can be stated as follows.

Lemma (Bihari)

Under the preceding assumptions on (1.32), *we have*

$$x(t) \le G^{-1}\left(G(M) + \int_{t_0}^{t} k(s)\,ds\right) \quad (1.33)$$

for all $t \ge t_0$, *such that both sides of* (1.33) *are defined. The function* $G(u)$ *is a primitive of* $1/g(x)$.

Proof. We shall denote

$$y(t) = M + \int_{t_0}^{t} k(s)g(x(s))\,ds$$

which implies $y(t_0) = M$ and

$$\dot{y}(t) = k(t)g(x(t)). \tag{1.34}$$

Inasmuch as $g(x)$ is a nondecreasing function and (1.32) is equivalent to $x(t) \le y(t)$, we get, from (1.34),

$$\dot{y}(t) \le k(t)g(y(t)), \qquad t \in [t_0, T). \tag{1.35}$$

This inequality is of the separable variables type, and we obtain easily

$$y(t) \le G^{-1}\left(G(M) + \int_{t_0}^{t} k(s)\,ds\right) \tag{1.36}$$

which implies (1.33).

We point out that the right-hand side of (1.33) could be defined only for $t \in [t_0, \tilde{T})$, with $\tilde{T} < T$. This situation occurs if the range of $G(u)$ does not contain all the values of $G(M) + \int_{t_0}^{t} k(s)\,ds$, for $t \in [t_0, T)$.

PROBLEMS

1. Consider the differential equation of Bernoulli type

$$\dot{x} = a(t)x + b(t)x^m$$

where m is a real number, $m \ne 0$ or 1, and $a(t)$, $b(t)$ are continuous real functions on some interval (α, β) of the real axis. Assume that $x = x(t)$ is a positive solution of the Bernoulli equation on (α, β). Prove that $y = x^{1-m}$ satisfies the linear equation

$$\dot{y} = (1 - m)a(t)y + (1 - m)b(t).$$

Derive from the preceding considerations a method of integrating the Bernoulli equation. Discuss the validity of this method with respect to the values of the parameter m.

2. Consider the differental equation (Riccati)

$$\dot{x} = a(t)x^2 + b(t)x + c(t)$$

where $a(t)$, $b(t)$, and $c(t)$ are continuous real functions on a given interval (α, β). Show that—generally—there exist solutions of the Riccati equation that are not defined on the whole interval (α, β). More precisely, such solutions tend to $\pm\infty$, as $t \to T \in (\alpha, \beta)$. Is T independent of the solution?

Assume that we know a solution $x = x_0(t)$ of the Riccati equation and show that by changing the unknown function by means of the formula $x = x_0(t) + 1/y$, we find a linear differential equation in y. Discuss the validity of the preceding procedure in reducing a Riccati equation to a linear one.

3. Show that any homogeneous differential equation $\dot{x} = f(x/t)$ can be reduced to an equation with separable variables by means of the change of the unknown function $x = ty$.
4. Consider the differential equation

$$M(t, x)\, dt + N(t, x)\, dx = 0$$

with $M(t, x)$ and $N(t, x)$ homogeneous functions of the same degree. Show that this equation is homogeneous in the sense considered in the preceding problem.
5. Let us consider the linear differential equation $\dot{x} = a(t)x + b(t)$, where $a(t)$ and $b(t)$ are continuous real functions on the half-axis $t \geq 0$. Prove the following statements:
   (a) If $a(t) \leq -m < 0$ and $b(t)$ is bounded on the half-axis $t \geq 0$, then any solution of the linear equation considered previously is bounded on the half-axis $t \geq 0$.
   (b) If $a(t) \geq m > 0$ and $b(t)$ is bounded, then there exists one and only one solution bounded on the half-axis $t \geq 0$. This solution is given by

$$x(t) = -\int_t^\infty b(s) \exp\left[\int_s^t a(u)\, du\right] ds.$$

   (c) If $\lim a(t) = -A$ and $\lim b(t) = B$ as $t \to \infty$, with $A > 0$, then any solution of the linear equation considered previously satisfies $\lim x(t) = B/A$ as $t \to \infty$.
   (d) If $A < 0$, there exists one and only one solution such that $\lim x(t)$ as $t \to \infty$ is finite.
   (e) When $a(t) = \text{const}$ and $b(t)$ is a periodic function with period $T > 0$, discuss the problem of existence of periodic solutions and express them by means of some formulas.
6. Find conditions under which the solutions of the linear equation $t\dot{x} = x + b(t)$ tend to a finite limit as $t \to 0+$.
7. Consider the differential equation $M(t, x)\, dt + N(t, x)\, dx = 0$ and assume that $(\partial M(t, x)/\partial x - \partial N(t, x)/\partial t)/N(t, x)$ is a function of t only. Prove that there exists an integrating factor that is a function of t only and find it.
8. Inasmuch as $x(t) = e^t$ is the unique solution of the linear equation $\dot{x} = x$, with the initial condition $x(0) = 1$, it follows that we can prove the properties of the exponential function starting from the preceding equation. For instance, $x(t + s) = x(t)x(s)$ for any real t and s. Indeed, both functions are solutions of the equation $\dot{x} = x$, for fixed s. Moreover, they reduce to $x(s)$ for $t = 0$. Therefore, they coincide for all real t, which means that $e^{t+s} = e^t \cdot e^s$. Prove other basic properties of the exponential function using the equation and the initial condition that defines it.
9. Let $x(t)$ be a continuous nonnegative function on $[t_0, T]$ and assume that

$$x(t) \leq M + \int_{t_0}^t x(s)\, d\sigma(s), \qquad t \in [t_0, T]$$

where M is a positive constant and $\sigma(t)$ is a nondecreasing function on $[t_0, T]$, such that $\sigma(t) = \sigma(t + 0)$. Prove that $x(t)$ satisfies

$$x(t) \leq M \exp[\sigma(t) - \sigma(t_0)].$$

Derive the Gronwall inequality from the preceding one. The integral inequality just considered is called the generalized Gronwall's inequality.

2

Existence Theory for Differential Equations

The aim of this chapter is to establish the main results concerning the existence and uniqueness of solutions for ordinary differential equations.

We shall use the classical method of successive approximations, which consists in the construction of a sequence of functions (scalar or vectorial) such that the limit of this sequence in the sense of uniform convergence is the solution of the given equation or system.

Another important tool we shall use in this chapter is the compactness criterion for families of continuous functions. This criterion, also called the Ascoli–Arzelà lemma, will be useful in some subsequent chapters.

§2.1. The First-Order Equation

Consider the first-order equation

$$\dot{x} = f(t, x) \tag{2.1}$$

where $f(t, x)$ is a continuous function of (t, x) in the rectangle

D:[1] $$|t - t_0| \leq a, \qquad |x - x_0| \leq b$$

with the initial condition

$$x(t_0) = x_0 . \tag{2.2}$$

[1] D represents *domain*, and is the notation that will be used throughout this text.

Assume that $x = x(t)$ is a solution of (2.1), satisfying also (2.2), defined in a neighborhood $I = \{t; |t - t_0| \leq \delta \leq a\}$ of $t = t_0$.

From (2.1) it follows that $\dot{x}(t)$ is a continuous function on the interval I, and, consequently, we can integrate both sides of (2.1) from t_0 to an arbitrary $t \in I$. One obtains

$$x(t) = x_0 + \int_{t_0}^{t} f(s, x(s))\, ds \tag{2.3}$$

if we take into account the initial condition (2.2). Thus, any solution of (2.1), (2.2) is also a solution of the integral equation (2.3).

Conversely, if $x = x(t)$ is a continuous solution of (2.3), defined on I, then we can take the derivatives of both sides in (2.3), because $f(t, x(t))$ is also continuous. This leads to (2.1). The condition (2.2) is obviously verified by $x(t)$ from (2.3).

We can express the result of the preceding discussion in the following manner.

Lemma

Consider differential equation (2.1) *with condition* (2.2) *and assume that* $f(t, x)$ *is continuous in* D.

Then, any solution $x = x(t)$ *of* (2.1), (2.2), *defined in an interval* $I = \{t; |t - t_0| \leq \delta \leq a\}$, *is a continuous solution of the integral equation* (2.3), *on the same interval.*

The converse statement is also true.

It is now obvious that we can prove the existence theorem for (2.3), instead of (2.1), (2.2). According to the method of successive approximations, let us consider the sequence $\{x_n(t)\}$, such that

$$x_n(t) = x_0 + \int_{t_0}^{t} f(s, x_{n-1}(s))\, ds, \qquad n = 1, 2, \ldots \tag{2.4}$$

where $x_0(t) \equiv x_0$, $t \in [t_0 - a, t_0 + a]$.

Theorem 2.1

Consider the differential equation (2.1) *and assume that*:

1. *the function* $f(t, x)$ *is continuous on* D;
2. *for any pair* (t, x), $(t, y) \in D$, *we have*

$$|f(t, x) - f(t, y)| \leq L\,|x - y| \tag{2.5}$$

where $L > 0$ *is a given constant.*

Then, there exists a unique solution $x = x(t)$ *of* (2.1), *satisfying the initial condition* (2.2) *and defined on the interval*

$$|t - t_0| \leq \delta = \min\left\{a, \frac{b}{M}\right\} \tag{2.6}$$

where $M = \sup|f(t, x)|$, $(t, x) \in D$.

Moreover, $x(t)$ *is the limit of the sequence* $\{x_n(t)\}$ *given by* (2.4), *uniformly on* $|t - t_0| \leq \delta$.

Proof. We shall prove that all the functions $x_n(t)$ are continuous on $|t - t_0| \leq \delta$ and that $x(t) = \lim_{n \to \infty} x_n(t)$ exists in the sense of uniform convergence. It will follow without any difficulty that $x(t)$ is a solution of (2.3)—that is, of (2.1), (2.2).

Because $x_0(t) \equiv x_0$, $t \in [t_0 - a, t_0 + a]$, by definition, from (2.4) it follows for $n = 1$:

$$x_1(t) = x_0 + \int_{t_0}^{t} f(s, x_0)\, ds \tag{2.7}$$

which proves that $x_1(t)$ is continuous on $|t - t_0| \leq a$, hence on $|t - t_0| \leq \delta$. Moreover, from (2.7) we obtain

$$|x_1(t) - x_0| \leq M|t - t_0| \leq M\delta \leq b \tag{2.8}$$

if $|t - t_0| \leq \delta$.

Assume now that $x_{n-1}(t)$ is continuous on $|t - t_0| \leq \delta$ and

$$|x_{n-1}(t) - x_0| \leq b$$

on this interval. From (2.4) we deduce that $x_n(t)$ is continuous on

$$|t - t_0| \leq \delta$$

and

$$|x_n(t) - x_0| \leq M|t - t_0| \leq M\delta \leq b. \tag{2.9}$$

Consequently, according to the principle of mathematical induction, we conclude that $\{x_n(t)\}$ consists in continuous functions on $|t - t_0| \leq \delta$ and their values belong to the interval $|x - x_0| \leq b$. In other words,

$$(t, x_n(t)) \in D, \qquad |t - t_0| \leq \delta, \qquad n = 1, 2, \ldots .$$

Let us prove now that the sequence $\{x_n(t)\}$ is uniformly convergent on $|t - t_0| \leq \delta$. For $n = 2$, Eq. (2.4) becomes

$$x_2(t) = x_0 + \int_{t_0}^{t} f(s, x_1(s))\, ds. \tag{2.10}$$

From (2.7) and (2.10) we get

$$|x_2(t) - x_1(t)| \le \left| \int_{t_0}^{t} |f(s, x_1(s)) - f(s, x_0)| \, ds \right|.$$

Taking into account condition (2.5), we obtain

$$|x_2(t) - x_1(t)| \le L \left| \int_{t_0}^{t} |x_1(s) - x_0| \, ds \right|. \tag{2.11}$$

Now, we can apply the first inequality (2.8) in the right-hand side of (2.11):

$$|x_2(t) - x_1(t)| \le ML \frac{|t - t_0|^2}{2!} \le ML \frac{\delta^2}{2!}. \tag{2.12}$$

To find an estimation for $|x_n(t) - x_{n-1}(t)|$, we shall again apply the method of induction. Starting from (2.8) and (2.12), assume that

$$|x_n(t) - x_{n-1}(t)| \le ML^{n-1} \frac{|t - t_0|^n}{n!} \le ML^{n-1} \frac{\delta^n}{n!} \tag{2.13}$$

and let us prove that such an inequality holds for $|x_{n+1}(t) - x_n(t)|$. Indeed, from (2.4) and

$$x_{n+1}(t) = x_0 + \int_{t_0}^{t} f(s, x_n(s)) \, ds$$

it follows that

$$|x_{n+1}(t) - x_n(t)| \le \left| \int_{t_0}^{t} |f(s, x_n(s)) - f(s, x_{n-1}(s))| \, ds \right|.$$

According to (2.5) and (2.13), we get

$$|x_{n+1}(t) - x_n(t)| \le ML^n \frac{|t - t_0|^{n+1}}{(n+1)!} \le ML^n \frac{\delta^{n+1}}{(n+1)!} \tag{2.14}$$

if we take into account that

$$\left| \int_{t_0}^{t} |s - t_0|^n \, ds \right| = \frac{|t - t_0|^{n+1}}{n+1}.$$

But (2.14) is the same as (2.13), with $n + 1$ instead of n. Consequently, inequality (2.13) holds for $n = 1, 2, \ldots$. We can also write

$$|x_n(t) - x_{n-1}(t)| \le \frac{M}{L} \frac{(L\delta)^n}{n!} \tag{2.15}$$

for $n = 1, 2, \ldots$ and $|t - t_0| \le \delta$.

Let us mention now that

$$x_n(t) = x_0 + [x_1(t) - x_0] + \cdots + [x_n(t) - x_{n-1}(t)]$$

which implies that the sequence $\{x_n(t)\}$ and the series

$$x_0 + \sum_{n=1}^{\infty} [x_n(t) - x_{n-1}(t)]$$

have the same convergence properties. From (2.15), according to the convergence criterion of Weierstrass, it follows that the series having the general term $x_n(t) - x_{n-1}(t)$ converges uniformly on $|t - t_0| \le \delta$. The same is true for the sequence $\{x_n(t)\}$, and denoting $x(t) = \lim_{n\to\infty} x_n(t)$, we can assert that

$$\lim_{n\to\infty} f(t, x_n(t)) = f(t, x(t)) \tag{2.16}$$

uniformly on $|t - t_0| \le \delta$. Equation (2.16) follows easily from

$$|f(t, x_n(t)) - f(t, x(t))| \le L\,|x_n(t) - x(t)|. \tag{2.17}$$

Taking (2.16) into account, from (2.4) we obtain, for $n \to \infty$,

$$x(t) = x_0 + \int_{t_0}^{t} f(s, x(s))\, ds$$

which is nothing else but Eq. (2.3). Consequently, there is at least one continuous solution of (2.3), or equivalently, of (2.1), (2.2).

We want to prove now that this solution is unique, that is, from

$$y(t) = x_0 + \int_{t_0}^{t} f(s, y(s))\, ds \tag{2.18}$$

on $|t - t_0| \le \delta$, it follows that $y(t) = x(t)$. Indeed, from (2.4) and (2.18), we obtain

$$|y(t) - x_n(t)| \le L \left| \int_{t_0}^{t} |y(s) - x_{n-1}(s)|\, ds \right|, \qquad n = 1, 2, \ldots. \tag{2.19}$$

But $|y(t) - x_0| \le b$ on $|t - t_0| \le \delta, y(t)$ being a solution of (2.18). It follows easily from (2.19) that

$$|y(t) - x_1(t)| \le bL\,|t - t_0| \tag{2.20}$$

on $|t - t_0| \le \delta$. Assume now

$$|y(t) - x_n(t)| \le bL^n \frac{|t - t_0|^n}{n!} \tag{2.21}$$

on the interval $|t - t_0| \le \delta$. From

$$|y(t) - x_{n+1}(t)| \le L \left| \int_{t_0}^{t} |y(s) - x_n(s)|\, ds \right|$$

and (2.21), one obtains

$$|y(t) - x_{n+1}(t)| \leq bL^{n+1} \frac{|t - t_0|^{n+1}}{(n+1)!}. \tag{2.22}$$

Consequently, (2.21) holds for any n, which leads to the conclusion that

$$y(t) = \lim_{n\to\infty} x_n(t) = x(t) \tag{2.23}$$

on the interval $|t - t_0| \leq \delta$. Thus is proved the uniqueness of the solution for (2.1), (2.2).

Remark 1. From (2.21) and (2.23), it follows that the error occurring in the method of successive approximations, if we stop at the nth step, is given by

$$|\varepsilon_n(x)| = |x(t) - x_n(t)| \leq b\frac{(L|t - t_0|)^n}{n!} \leq b\frac{(L\delta)^n}{n!}. \tag{2.24}$$

Remark 2. The condition (2.5) is frequently called the *Lipschitz condition* for $f(t, x)$, with respect to x.

It is easy to see that this condition takes place for $f(t, x)$ if there exists the derivative $\partial f/\partial x$ and it is bounded on D. Indeed, we can write

$$f(t, x) - f(t, y) = \left(\frac{\partial f}{\partial x}\right)^* (x - y) \tag{2.25}$$

where the asterisk indicates the fact that $\partial f/\partial x$ is taken at the point (t, ξ), with ξ between x and y. From (2.25) we can derive an inequality of the form (2.5).

Remark 3. Assume that D is replaced by the strip

$$D': \qquad |t - t_0| \leq a, \qquad |x| < +\infty.$$

Then $\{x_n(t)\}$ consists of continuous functions on $|t - t_0| \leq a$ and converges uniformly on the same interval. Consequently, the solution $x(t)$ of (2.1), (2.2) is defined and unique on the same interval. Of course, we assume that (2.5) holds in the strip D'.

§2.2. First-Order Differential Systems

Consider now the differential system

$$\dot{x}_i = f_i(t, x_1, \ldots, x_n), \qquad i = 1, 2, \ldots, n \tag{2.26}$$

with the initial conditions

$$x_i(t_0) = x_i^{(0)}, \qquad i = 1, 2, \ldots, n. \tag{2.27}$$

Before considering the problem of existence and uniqueness of the solution, we shall introduce *vector* notation.

Let $x = (x_1, \ldots, x_n)$ be the n vector the coordinates of which are the numbers $x_1, x_2, \ldots, x_n$. It is well known that the n vectors form a linear space over the field of real (or complex) numbers, with respect to the basic operations defined as follows:

1. If $x = (x_1, \ldots, x_n)$, $y = (y_1, \ldots, y_n)$, then

$$x + y = (x_1 + y_1, \ldots, x_n + y_n).$$

2. If $x = (x_1, \ldots, x_n)$ and c is a scalar, then

$$cx = (cx_1, \ldots, cx_n).$$

For real vectors, the norm is given by

$$\|x\| = (x_1{}^2 + \cdots + x_n{}^2)^{1/2}$$

and the following properties are fundamental:

(a) $\|x\| \geq 0$, the equality being possible only for $x = \theta = (0, \ldots, 0)$.
(b) $\|x + y\| \leq \|x\| + \|y\|$, for any pair x, y.
(c) $\|cx\| = |c| \, \|x\|$, where c is a scalar and x an arbitrary vector.
(d) For any $x = (x_1, \ldots, x_n)$, we have $\max_j |x_j| \leq \|x\| \leq n \max_j |x_j|$ and

$$\frac{1}{n} \sum_{j=1}^{n} |x_j| \leq \|x\| \leq \sum_{j=1}^{n} |x_j| .$$

A *vector function* $x = x(t) = (x_1(t), \ldots, x_n(t))$ is completely determined by the set of n scalar functions—its coordinates. It is possible to extend the concepts of the classical analysis (i.e., the concept related to scalar functions) to vector functions. For instance, we shall say that $\lim_{t \to t_0} x(t) = \xi$, if and only if $\lim_{t \to t_0} \|x(t) - \xi\| = 0$. From the inequalities

$$\max_j |x_j(t) - \xi_j| \leq \|x(t) - \xi\| \leq \sum_{j=1}^{n} |x_j(t) - \xi_j|$$

it follows that $x(t) \to \xi$ for $t \to t_0$, if and only if $x_j(t) \to \xi_j$ for $t \to t_0$, $j = 1, 2, \ldots, n$. It is obvious that a vector function $x = x(t)$ is continuous at $t = t_0$ if and only if its components are continuous at the same point.

The derivative of a vector function will be defined by

$$\dot{x}(t) = \lim_{h \to 0} \frac{x(t + h) - x(t)}{h}$$

if the limit exists. According to the preceding remark, this case occurs if and only if

$$\lim_{h \to 0} \frac{x_j(t + h) - x_j(t)}{h}$$

exists for any j, $j = 1, 2, \ldots, n$. We get

$$\dot{x}(t) = (\dot{x}_1(t), \ldots, \dot{x}_n(t)).$$

Concerning the integral of a continuous vector function, elementary considerations lead to the following formulas:

$$\int_a^b x(t)\,dt = \lim \sum_1^m x(\tau_k)(t_k - t_{k-1})$$
$$= \left(\int_a^b x_1(t)\,dt, \ldots, \int_a^b x_n(t)\,dt\right).$$

An inequality concerning the integral of a vector function plays an important role in the subsequent considerations. Namely,

$$\left\| \int_a^b x(t)\,dt \right\| \leq \int_a^b \|x(t)\|\,dt \tag{2.28}$$

where $x(t)$ is continuous on $[a, b]$, $a < b$. Inequality (2.28) follows easily from

$$\left\| \sum_1^m x(\tau_k)(t_k - t_{k-1}) \right\| \leq \sum_1^m \|x(\tau_k)\|\,(t_k - t_{k-1})$$

if we take the limit of both sides in the sense required by the definition of the integral. Obviously, similar considerations hold for vector functions of several variables.

Now we can go back to our system (2.1). If we denote

$$f(t, x) = (f_1(t, x_1, \ldots, x_n), \ldots, f_n(t, x_1, \ldots, x_n)) \tag{2.29}$$

then (2.26) can be written

$$\dot{x} = f(t, x) \tag{2.30}$$

that is, in a form identical to that corresponding to the scalar case, with the single difference that x and f are now vectors (more precisely, vector functions).

The initial conditions (2.27) are equivalent to

$$x(t_0) = x^0 \tag{2.31}$$

where x^0 is a given n vector (we shall use superscripts for vectors to avoid the confusion which might result from the fact that subscripts are used to designate the components of the vector). We can now formulate the existence and uniqueness theorem for systems.

Theorem 2.2

Assume the following conditions are satisfied:

1. *The vector function $f(t, x)$ is continuous on the set*

$$\Delta: \qquad |t - t_0| \leq a, \qquad \|x - x^0\| \leq b.$$

2. *$f(t, x)$ satisfies the Lipschitz condition with respect to x*

$$\|f(t, x) - f(t, y)\| \leq L\,\|x - y\|. \tag{2.32}$$

Then, there exists a unique solution $x = x(t)$ of the system (2.30), *satisfying also the initial condition* (2.31). *This solution is defined on the interval*

$$|t - t_0| \leq \delta = \min\left\{a, \frac{b}{M}\right\} \tag{2.33}$$

where $M = \sup\|f(t, x)\|$, $(t, x) \in \Delta$. Moreover, the solution is the uniform limit on $|t - t_0| \leq \delta$ of the sequence of vector functions $\{x^m(t)\}$, where $x^0(t) \equiv x^0$ and

$$x^m(t) = x^0 + \int_{t_0}^{t} f(s, x^{m-1}(s))\,ds, \qquad m = 1, 2, \ldots.$$

The proof of this theorem follows the same lines as the proof of Theorem 2.1. We have to take into account that x and f are vectors and, consequently, the norm should be used instead of the absolute value. The reader will carry out the proof of Theorem 2.2 himself.

Remark. It is easy to see that Lipschitz condition (2.32) is equivalent to

$$|f_i(t, x_1, \ldots, x_n) - f_i(t, y_1, \ldots, y_n)| \leq L_1 \sum_{j=1}^{n} |x_j - y_j|, \qquad i = 1, 2, \ldots, n \tag{2.34}$$

for a convenient L_1. We have to take into account the property of the norm expressed by (d).

Another equivalent form of condition (2.32) is

$$|f_i(t, x_1, \ldots, x_n) - f_i(t, y_1, \ldots, y_n)| \leq L_2 \max_j |x_j - y_j|, \qquad i = 1, 2, \ldots, n. \tag{2.35}$$

Of course, the existence and boundedness in Δ of the derivatives $\partial f_i/\partial x_j$ imply condition (2.32).

§2.3. Equations and Systems of Higher Order

Theorem 2.2 is more general than it could appear at first glance. Actually, the equations and systems of higher order can be reduced by elementary transforms to the form (2.26) in §2.2.

Let us begin with the nth-order equation

$$x^{(n)} = f(t, x, \dot{x}, \ldots, x^{(n-1)}) \tag{2.36}$$

where $f(t, x, x_1, \ldots, x_{n-1})$ is continuous in the set

$\Gamma\colon |t - t_0| \le a, |x - x_0| \le b, |x_1 - x_{1,0}| \le b, \ldots, |x_{n-1} - x_{n-1,0}| \le b$

and satisfies also the Lipschitz condition

$$|f(t, x, x_1, \ldots, x_{n-1}) - f(t, y, y_1, \ldots, y_{n-1})| \\ \le L(|x - y| + |x_1 - y_1| + \cdots + |x_{n-1} - y_{n-1}|). \tag{2.37}$$

It is very easy to reduce Eq. (2.36) to a first-order system of the form (2.26) in §2.2. Indeed, let us denote $\dot{x} = x_1$, $\ddot{x} = \dot{x}_1 = x_2, \ldots, x^{(n-1)} = \dot{x}_{n-2} = x_{n-1}$. Then (2.36) becomes

$$\dot{x}_{n-1} = f(t, x, x_1, \ldots, x_{n-1}).$$

Consequently, if $x = x(t)$ is a solution of (2.1) on a given interval $|t - t_0| \le \delta$, then

$$x = x(t), x_1 = \dot{x}(t), \ldots, x_{n-1} = x^{(n-1)}(t) \tag{2.38}$$

is a solution of the system

$$\begin{cases} \dot{x} = x_1, \\ \dot{x}_1 = x_2, \\ \quad\vdots \\ \dot{x}_{n-2} = x_{n-1}, \\ \dot{x}_{n-1} = f(t, x, x_1, \ldots, x_{n-1}), \end{cases} \tag{2.39}$$

that is, a system of the form (2.26) in §2.2.

Conversely, assume we have a solution $(x(t), x_1(t), \ldots, x_{n-1}(t))$ of system (2.39), defined on a given interval $|t - t_0| \le \delta$. Then $x = x(t)$ is a solution of Eq. (2.36). To prove this statement, it is sufficient to point out that $x = x(t)$ has derivatives of order $\le n$ [because from (2.39) it follows $x_2 = \dot{x}_1 = \ddot{x}$, $x_3 = \dot{x}_2 = \dddot{x}, \ldots$] and to eliminate $x_1, x_2, \ldots, x_{n-1}$ from Eqs. (2.39). One obtains Eq. (2.36). Formula (2.38) establishes the connection between the solutions of Eq. (2.36) and system (2.39), which leads to the following initial conditions for (2.36):

$$x(t_0) = x_0, \dot{x}(t_0) = x_{1,0}, \ldots, x^{(n-1)}(t_0) = x_{n-1,0}. \tag{2.40}$$

Taking Theorem 2.2 into account, we can state the following existence and uniqueness theorem for (2.36), (2.40).

Theorem 2.3

Consider Eq. (2.36) *with initial conditions* (2.40) *and assume that* $f(t, x, x_1, \ldots, x_{n-1})$ *is a continuous function on* Γ, *satisfying Lipschitz condition* (2.37). *Then there exists a unique solution $x = x(t)$ of this problem, defined on the interval $|t - t_0| \leq \delta \leq a$, with δ sufficiently small.*

It is interesting to point out that the procedure just used is such that for any linear equation

$$x^{(n)} = a_1(t)x^{(n-1)} + a_2(t)x^{(n-2)} + \cdots + a_n(t)x + b(t)$$

the corresponding system is also linear:

$$\begin{aligned} \dot{x} &= x_1, \\ \dot{x}_1 &= x_2, \\ &\vdots \\ \dot{x}_{n-2} &= x_{n-1}, \\ \dot{x}_{n-1} &= a_1(t)x_{n-1} + a_2(t)x_{n-2} + \cdots + a_n(t)x + b(t). \end{aligned}$$

Finally, let us remark that the character of an equation of being, or not being, homogeneous [i.e., $b(t) \equiv 0$ or $b(t) \not\equiv 0$] is also preserved by transformation.

The case of differential systems of higher order does not differ essentially from the preceding one. Assume, for the sake of simplicity, that we have a system containing two unknown functions, x and y:

$$\begin{cases} x^{(n)} = f(t, x, \dot{x}, \ldots, x^{(n-1)}, y, \dot{y}, \ldots, y^{(m-1)}) \\ y^{(m)} = g(t, x, \dot{x}, \ldots, x^{(n-1)}, y, \dot{y}, \ldots, y^{(m-1)}). \end{cases} \tag{2.41}$$

System (2.41) is obviously equivalent to the following system of first-order equations:

$$\begin{aligned} \dot{x} &= x_1, \\ \dot{x}_1 &= x_2, \\ &\vdots \\ \dot{x}_{n-1} &= f(t, x, x_1, \ldots, x_{n-1}, y, y_1, \ldots, y_{m-1}), \\ \dot{y} &= y_1, \\ \dot{y}_1 &= y_2, \\ &\vdots \\ \dot{y}_{m-1} &= g(t, x, x_1, \ldots, x_{n-1}, y, y_1, \ldots, y_{m-1}). \end{aligned}$$

The initial conditions should have the form

$$\begin{aligned} x(t_0) = x_0, \dot{x}(t_0) = x_{1,0}, \ldots, x^{(n-1)}(t_0) = x_{n-1,0} \\ y(t_0) = y_0, \dot{y}(t_0) = y_{1,0}, \ldots, y^{(m-1)}(t_0) = y_{m-1,0}. \end{aligned} \tag{2.42}$$

We leave for the reader the task of formulating the corresponding existence and uniqueness theorem for (2.41), (2.42), analogous to Theorem

2.3. Similar considerations can be made for differential systems of higher order with more than two unknown functions.

§2.4. Peano Existence Theorem

The Lipschitz condition we assumed in the preceding theorems is, essentially, related to the uniqueness property and the method of proof. It is not necessary for the existence of the solution. Clarified by Peano, this fact led to an essentially new proof of the existence of solutions.

Before proceeding further with Peano's theorem, we shall give a compactness criterion for families of continuous functions on a given interval $I = [a, b]$.

Lemma (Ascoli–Arzelà)

Let $\mathscr{F} = \{f(t)\}$ be an infinite family of (real) continuous functions on I. Assume that:

1. *$\mathscr{F}$ is uniformly bounded on I—that is, there exists $M > 0$ such that $|f(t)| \le M$, for any $t \in I$ and any $f \in \mathscr{F}$;*
2. *$\mathscr{F}$ is equicontinuous on I—that is, for any $\varepsilon > 0$, there exists $\delta > 0$, such that*

$$|f(t) - f(s)| < \varepsilon \quad \text{for} \quad |t - s| < \delta, \qquad f \in \mathscr{F}.$$

Then $\mathscr{F}$ is relatively compact in the sense of uniform convergence on I—that is, any infinite part of $\mathscr{F}$ contains a sequence that is uniformly convergent on I.

Proof. Let us consider the set $\{r_n\}$ of all rational numbers in $[a, b]$. It is well known that this set is denumerable and, consequently, we can use the sequence notation.

Assume now that $\mathscr{G}$ is an infinite part of $\mathscr{F}$. Without loss of generality, we can assume that $\mathscr{G}$ is a sequence of functions: $\mathscr{G} = \{f_n(t)\}$. Now let us consider the sequence of numbers $\{f_n(r_1)\}$. We have $|f_n(r_1)| \le M$, and from this sequence (which is bounded!) we can extract a convergent subsequence: say, $\{f_{1n}(r_1)\}$. It is known that the sequence of functions $\{f_{1n}(t)\}$ is convergent at $t = r_1$.

Applying the same procedure as before to $\{f_{1n}(t)\}$, we can find a subsequence—say, $\{f_{2n}(t)\}$—that converges at $t = r_2$. As a subsequence of $\{f_{1n}(t)\}$, $\{f_{2n}(t)\}$ converges also at $t = r_1$. We can proceed further in the same way and obtain the sequences $\{f_{kn}(t)\}$, $k > 2$, with the following properties:

(a) $\{f_{kn}(t)\}$ is a subsequence of $\{f_{hn}(t)\}$, whenever $h < k$;
(b) $\{f_{kn}(t)\}$ is convergent at $t = r_j, j = 1, 2, \ldots, k$.

Let us consider now the sequence of functions $\{f_{nn}(t)\}$. Excepting a finite number of terms, it is a subsequence of any sequence $\{f_{kn}(t)\}$, $k = 1, 2, \ldots$. This follows easily if we take into account the properties (a) and (b). Hence, $\{f_{nn}(t)\}$ converges at $t = r_k$, $k = 1, 2, \ldots$; in other words, it converges for any rational $t \in I$.

We shall prove that $\{f_{nn}(t)\}$ is uniformly convergent on I. Indeed let $\varepsilon > 0$ be arbitrary and consider the corresponding number $\delta = \delta(\varepsilon/3)$. Consider also a division $a = t_0 < t_1 < \cdots < t_p = b$ of I, such that $\max(t_q - t_{q-1}) < \delta$. We can obviously assume that all t_q, $1 \le q < p$, are rational. It is clear that for any $t \in I$, there exists at least one t_q, such that $|t - t_q| < \delta$. But

$$|f_{nn}(t) - f_{mm}(t)| \le |f_{nn}(t) - f_{nn}(t_q)| + |f_{nn}(t_q) - f_{mm}(t_q)| + |f_{mm}(t_q) - f_{mm}(t)| < \frac{\varepsilon}{3} + \frac{\varepsilon}{3} + \frac{\varepsilon}{3} = \varepsilon$$

whenever $n, m \ge N(\varepsilon)$. Indeed, from $|t - t_q| < \delta$, it follows that

$$|f_{nn}(t) - f_{nn}(t_q)| < \frac{\varepsilon}{3} \quad \text{and} \quad |f_{mm}(t_q) - f_{mm}(t)| < \frac{\varepsilon}{3}$$

for all n, m. On the other hand, the sequences $\{f_{nn}(t_q)\}$, $1 \le q < p$, are convergent, and we can write

$$|f_{nn}(t_q) - f_{mm}(t_q)| < \frac{\varepsilon}{3}, \qquad 1 \le q < p$$

for $n, m \ge N(\varepsilon)$. Summing up the last considerations, we get

$$|f_{nn}(t) - f_{mm}(t)| < \varepsilon, \qquad t \in I, \qquad n, m \ge N(\varepsilon)$$

which means that $\{f_{nn}(t)\}$ is a uniformly convergent sequence. The lemma is thus proved.

Corollary

Let $\mathscr{F} = \{f(t)\}$ be an infinite set of continuous vector functions from $I = [a, b]$ into R^n, satisfying the following conditions:

1. *It is uniformly bounded on I—that is, $\|f(t)\| \le M$, for any $t \in I$, and $f \in \mathscr{F}$, for some $M > 0$;*
2. *It is equicontinuous on I—that is, for every $\varepsilon > 0$ we can find $\delta > 0$, such that*

$$\|f(t) - f(s)\| < \varepsilon \quad \text{for} \quad |t - s| < \delta, \qquad f \in \mathscr{F}.$$

Then $\mathscr{F}$ is relatively compact in the sense of uniform convergence on I.

The proof will follow immediately from the fact that the sets $\mathscr{F}_k = \{f_k(t)\}$, where $f_k(t)$ denotes the kth component of $f(t)$, $k = 1, 2, \dots, n$, are uniformly bounded and equicontinuous on I.

Indeed, let $\{f^p\} \in \mathscr{F}$ be an arbitrary sequence. Since $\{f_1{}^p\} \in \mathscr{F}_1$, there then exists a subsequence $\{f_1{}^{p'}\}$ that converges uniformly on I. Consider now the sequence $\{f_2{}^{p'}\} \in \mathscr{F}_2$. It contains a subsequence $\{f_2{}^{p''}\}$ that is uniformly convergent on I, and so on.

Consider now the scalar differential equation

$$\dot{x} = f(t, x) \tag{2.43}$$

with $f(t, x)$ continuous on the rectangle

$$D: \qquad |t - t_0| \le a, \qquad |x - x_0| \le b.$$

An important concept related to such an equation (or system) is that of *approximate solution*. More precisely, we shall say that $x = x(t)$, given on $|t - t_0| \le \delta \le a$ and such that $|x(t) - x_0| \le b$ is an ε-approximate solution of (2.43), corresponding to the initial condition

$$x(t_0) = x_0 \tag{2.44}$$

if the following properties hold:

1. $x(t)$ is continuous on $|t - t_0| \le \delta$;
2. The derivative $\dot{x}(t)$ exists, except at a finite number of points, and it is piecewise continuous;
3. For all t for which $\dot{x}(t)$ is defined we have

$$|\dot{x}(t) - f(t, x(t))| < \varepsilon. \tag{2.45}$$

We shall now prove another lemma showing that for any $\varepsilon > 0$ there exists at least one ε-approximate solution of (2.43), (2.44).

Lemma (existence of approximate solutions)

Consider Eq. (2.43) *with continuous* $f(t, x)$ *on the rectangle* D, *and let* $\varepsilon > 0$ *be arbitrary. Then there exists at least one* ε*-approximate solution of* (2.43), (2.44), *defined on*

$$|t - t_0| \le \delta = \min\left\{a, \frac{b}{M}\right\} \tag{2.46}$$

where $M = \sup|f(t, x)|$, $(t, x) \in D$.

Proof. Inasmuch as $f(t, x)$ is a continuous function on the compact set D, it follows that $f(t, x)$ is uniformly continuous. Consequently, we can find

$\delta(\varepsilon) > 0$, such that $|f(t, x) - f(s, y)| < \varepsilon$, for $(t, x), (s, y) \in D$, $|t - s| < \delta$, $|x - y| < \delta$.

We shall construct now the approximate solution for $t \in [t_0, t_0 + \delta]$, the construction being completely similar for $t \in [t_0 - \delta, t_0]$. Let us consider a division

$$t_0 < t_1 < \cdots < t_n = t_0 + \delta$$

of $[t_0, t_0 + \delta]$, such that

$$\max_k (t_k - t_{k-1}) < \alpha = \min\left\{\delta, \frac{\delta}{M}\right\}. \tag{2.47}$$

We define the function $x(t)$ as follows:

$$x(t_0) = x_0, \qquad x(t) = x(t_k) + f(t_k, x(t_k)) \cdot (t - t_k) \tag{2.48}$$

on $t_k < t \leq t_{k+1}$, $k = 0, 1, \ldots, n - 1$. It is obvious that $x(t)$ satisfies the first two properties from the definition of an ε-approximate solution. We want to prove now that the last property is also fulfilled. Indeed, $\dot{x}(t) = f(t_k, x(t_k))$ on (t_k, t_{k+1}), and

$$|\dot{x}(t) - f(t, x(t))| = |f(t_k, x(t_k)) - f(t, x(t))| < \varepsilon$$

because $|t - t_k| < \alpha \leq \delta$, $|x(t) - x(t_k)| = |f(t_k, x(t_k))| \cdot |t - t_k| < M\alpha \leq \delta$. Thus, (2.45) is verified and the lemma is proved.

Remark. The approximate solution $x(t)$ satisfies on $|t - t_0| \leq \delta$ the Lipschitz condition

$$|x(t) - x(s)| \leq M|t - s|. \tag{2.49}$$

Inequality (2.7) is trivial when t and s belong to the same (t_k, t_{k+1}). In the general case, it follows from a very simple geometric argument. Namely, it suffices to remark that the slope of the line passing by $(t, x(t))$ and $(s, x(s))$ cannot exceed, in absolute value, the maximum of slopes of the segments in the broken line which is the graph of $x(t)$. We can now supply the proof of the Peano existence theorem.

Theorem 2.4

Consider Eq. (2.43) *with $f(t, x)$ continuous on D. Then there exists at least one solution of* (2.43), (2.44), *defined on the interval given by* (2.46).

Proof. Let $\{\varepsilon_n\}$ be a sequence of positive numbers, such that $\varepsilon_n \to 0$ as $n \to \infty$. From the preceding lemma, it follows the existence of an ε_n-approximate solution—say, $x_n(t)$—for every natural n. Consider the sequence

$\{x_n(t)\}$, the terms of which are continuous functions on the interval (2.46), and remark that $|x_n(t) - x_0| \le b$, which means that $\{x_n(t)\}$ is uniformly bounded. Moreover, the sequence is equicontinuous, as can be seen easily from the remark to the preceding lemma. Consequently, there exists a subsequence—let us denote it again by $\{x_n(t)\}$—such that $\lim x_n(t) = x(t)$ uniformly on (2.46) as $n \to \infty$.

On the other hand, we denote

$$\Delta_n(t) = \dot{x}_n(t) - f(t, x_n(t)) \tag{2.50}$$

for all t such that the right-hand side of (2.50) makes sense and $\Delta_n(t) = 0$ for all other t in (2.46). We have $\Delta_n(t) \to 0$ as $n \to \infty$, uniformly on (2.46), because $|\Delta_n(t)| < \varepsilon_n$ on (2.46). From (2.50) we get

$$x_n(t) = x_0 + \int_{t_0}^{t} f(s, x_n(s))\, ds + \int_{t_0}^{t} \Delta_n(s)\, ds \tag{2.51}$$

taking into account (2.44) for $x_n(t)$.

But $f(t, x_n(t)) \to f(t, x(t))$ as $n \to \infty$, uniformly on the interval (2.46). Indeed, $f(t, x)$ is uniformly continuous on D, and for any $\varepsilon > 0$ we can find $\eta > 0$ such that $|f(t, x) - f(t, y)| < \varepsilon$ if $|x - y| < \eta$. Consequently, for any $\varepsilon > 0$, we shall have $|f(t, x_n(t)) - f(t, x(t)| < \varepsilon$ if $|x_n(t) - x(t)| < \eta$. But the last inequality holds for sufficiently large n, independently with respect to t in the interval (2.46).

Now, we can take the limit in both sides of (2.51) and obtain

$$x(t) = x_0 + \int_{t_0}^{t} f(s, x(s))\, ds \tag{2.52}$$

because

$$\left| \int_{t_0}^{t} \Delta_n(s)\, ds \right| \le \varepsilon_n |t - t_0| \le \varepsilon_n \delta$$

and $\varepsilon_n \to 0$ as $n \to \infty$. Therefore, $x(t)$ is a solution of (2.43), (2.44) on the interval (2.46), as shown in the lemma from §2.1.

§2.5. A Uniqueness Theorem

Inasmuch as we know that continuity is sufficient for the existence of at least one solution, and, on the other hand, it is not sufficient for the uniqueness (see §1.4), one can ask what condition we have to assume (more than continuity) to ensure the uniqueness of the initial-value problem (2.43), (2.44). We shall prove here a uniqueness theorem, which is a special case of a result of Osgood. Our tool will be the lemma of Bihari (§1.5).

Theorem 2.5

Consider the differential equation

$$\dot{x} = f(t, x) \tag{2.53}$$

with the initial condition

$$x(t_0) = x_0 \tag{2.54}$$

where $f(t, x)$ is continuous on

D: $$|t - t_0| \le a, \qquad |x - x_0| \le b$$

and such that

$$|f(t, x) - f(t, y)| \le \varphi(|x - y|). \tag{2.55}$$

If $\varphi(u)$ is a continuous nondecreasing function on $0 < u \le A$, with $\varphi(0) = 0$ and

$$\int_0^A \frac{du}{\varphi(u)} = +\infty \tag{2.56}$$

then the solution of problem (2.53), (2.54) *is unique.*

Proof. We shall consider only the case $t \ge t_0$. The case $t \le t_0$ reduces to the preceding one for the equation $\dot{x} = -f(-t, x)$.

Assume that there exist two solutions $x(t)$ and $y(t)$ of our problem, both defined in a neighborhood at the right of t_0. We have

$$x(t) = x_0 + \int_{t_0}^t f(s, x(s))\, ds$$

$$y(t) = y_0 + \int_{t_0}^t f(s, y(s))\, ds$$

which lead easily to

$$|x(t) - y(t)| \le \int_{t_0}^t |f(s, x(s)) - f(s, y(s))|\, ds$$

or

$$|x(t) - y(t)| < \varepsilon + \int_{t_0}^t \varphi(|x(s) - y(s)|)\, ds, \qquad \varepsilon > 0. \tag{2.57}$$

We can now apply Bihari's lemma to (2.57), getting

$$|x(t) - y(t)| < \Phi^{-1}(\Phi(\varepsilon) + t - t_0) \tag{2.58}$$

for those $t \ge t_0$ for which the right-hand side is defined. Here $\Phi(u)$ is a primitive for $1/\varphi(u)$.

We shall prove that the right-hand side of (2.58) tends toward zero as $\varepsilon \to 0$. Inasmuch as $|x(t) - y(t)|$ is independent of ε, it follows that $x(t) \equiv y(t)$, which we need. Let us remark that condition (2.56) implies $\Phi(\varepsilon) \to -\infty$ for $\varepsilon \to 0$, no matter how we choose the primitive of $1/\varphi(u)$. Thus $\Phi^{-1}(u) \to 0$ as $u \to -\infty$. Consequently, when $\varepsilon \to 0$ in the inequality (2.58), the right-hand side tends toward zero (for all finite t!). Theorem 2.5 is thus proved.

Remark 1. The condition on $\varphi(u)$ to be nondecreasing is not necessary, as follows from Osgood's proof.

Remark 2. The Lipschitz condition corresponds to $\varphi(u) = Lu$. Another possible choice for $\varphi(u)$ is, for instance,

$$\varphi(u) = Lu\,|\ln u|.$$

PROBLEMS

1. Consider the complex differential equation $dw/dz = f(z, w)$, where $f(z, w)$ is continuous for $|z - z_0| \le a$, $|w - w_0| \le b$, and holomorphic inside. Let M be a positive number such that $|f(z, w)| \le M$ in the domain of definition and denote $\delta = \min\{a, b/M\}$. Prove by successive approximations method that there exists a unique holomorphic function $w = w(z)$, $w(z_0) = w_0$, satisfying the given equation for all z such that $|z - z_0| < \delta$.
Hint: Consider the integral equation

$$w(z) = w_0 + \int_{z_0}^{z} f(\zeta, w(\zeta))\, d\zeta$$

where the complex integral is taken along the segment joining z_0 and z. Notice that our assumptions imply a Lipschitz condition for $f(z, w)$ with respect to w and then follow the usual pattern.

2. Consider the complex equation $dw/dz = f(w)$, with

$$f(w) = M\left[\frac{1}{2}\left(1 + \frac{w}{b}\right)\right]^{1/n}$$

taking the principal value in the right-hand side. Find the solution satisfying the initial condition $w(0) = 0$ and prove that the best result concerning the estimate of the radius of holomorphy of the solution is that already given: $\delta = \min\{a, b/M\}$.

3. Let us consider the differential system depending on parameters $\dot{x} = f(t, x; \lambda)$. Assume that the right member is a continuous vector function on $|t - t_0| \le a$, $\|x - x_0\| \le b$, $\lambda \in \Lambda \subset R^m$, satisfying a Lipschitz condition with respect to x. Moreover, assume that Λ is a compact set in R^m. Denote by $x = x(t; \lambda)$ the solution of the system, such that $x(t_0; \lambda) = x_0$. Investigate the function $x(t; \lambda)$. In particular, find conditions under which this function is differentiable with respect to λ.

4. Prove that the solution of the differential equation $x^{(n)} = f(t)$, with the usual initial conditions for an nth-order equation, is given by the formula

$$x(t) = x_0 + (t - t_0)x_0' + \cdots + \frac{(t-t_0)^{n-1}}{(n-1)!} x_0^{(n-1)} + \int_{t_0}^{t} \frac{(t-s)^{n-1}}{(n-1)!} f(s)\, ds.$$

Use this result to transform the general nth-order differential equation $x^{(n)} = f(t, x, \dot{x}, \ldots, x^{(n-1)})$ into an integro-differential equation, and then prove directly the existence and uniqueness theorem.

5. Consider the differential equation $F(t, x, \dot{x}) = 0$ where the real valued function $F(t, x, y)$ is continuous on

$$D: \qquad |t - t_0| \le a, \qquad |x - x_0| \le b, \qquad |y - y_0| \le c.$$

Assume further that $F(t_0, x_0, y_0) = 0$ and that there exist two numbers $\mu \ne 0$, $0 \le L < 1$, such that $|y - y_0 + \mu F(t, x, y)| \le c$ on D and

$$|y - z + \mu[F(t, x, y) - F(t, x, z)]| \le L|y - z|$$

for $(t, x, y), (t, x, z) \in D$. Prove the existence of at least one solution $x = x(t)$ of the given equation, such that $x(t_0) = x_0$.
Hint: Use the integro-differential equation

$$x(t) = x_0 + \int_{t_0}^{t} \{\dot{x}(s) + \mu F(s, x(s), \dot{x}(s))\}\, ds.$$

6. Prove the uniqueness theorem for scalar differential equations of the form $\dot{x} = f(t, x)$, with the initial condition $x(t_0) = x_0$, when $f(t, x)$ is continuous on $|t - t_0| \le a$, $|x - x_0| \le b$ and satisfies a condition of the Osgood type but without the assumption that $\varphi(u)$ (see Theorem 2.5) is nondecreasing.

7. Consider the scalar differential equation $\dot{x} = f(t, x)$, with $f(t, x)$ continuous on $t_0 \le t \le t_0 + a$, $|x - x_0| \le b$, and such that $|f(t, x) - f(t, y)|(t - t_0) \le k|x - y|$. If $0 < k \le 1$, then uniqueness holds at (t_0, x_0).
Hint: Assume that there exist at least two solutions through (t_0, x_0)—say, $x(t)$ and $y(t)$—both defined on $[t_0, t_0 + \delta]$. Consider the function $F(t) = [x(t) - y(t)]/(t - t_0)$, and use the l'Hospital rule to prove that it is continuous on $[t_0, t_0 + \delta]$. Then prove that $F(t)$ is identically zero on $[t_0, t_0 + \delta]$.

8. Consider the differential system $\dot{x} = f(t, x)$, $x, f \subset R^n$, with $f(t, x)$ continuous on $[t_0, t_0 + a]$, $\|x - x^0\| \le b$. Assume further that $f(t, x)$ satisfies a Lipschitz condition with respect to x, the corresponding constant being denoted by L. If $x(t)$ denotes the solution through (t_0, x^0) and $y(t)$ is an approximate solution through the same point, find an estimate for the error $x(t) - y(t)$.
Hint: First, find an integral inequality of the Gronwall type for the error. One starts from

$$x(t) = x^0 + \int_{t_0}^{t} f(s, x(s))\, ds,$$

$$y(t) = x^0 + \int_{t_0}^{t} f(s, y(s))\, ds + \int_{t_0}^{t} \Delta(s)\, ds, \qquad |\Delta(t)| < \varepsilon.$$

3

Some Global Problems for Ordinary Differential Equations

An inspection of the results included in Chapter 2 shows that they are all concerned with properties of solutions holding in a neighborhood of the initial point (t_0, x^0). This neighborhood is generally smaller than the domain of definition of the right member of the differential equation (or system), a feature suggesting that further arguments are necessary to elucidate the existence, the uniqueness, and other related problems.

The aim of this chapter is the investigation of such problems as global uniqueness, the continuation of solutions, the continuous dependence of solutions with respect to their initial values, and so on. A general result on differential inequalities will also be proved and utilized in obtaining analytic criteria of global existence.

§3.1. Statement of the Problems

Let us consider the differential system

$$\dot{x} = f(t, x) \tag{3.1}$$

where x and f are n-vectors, the last one being defined in the domain Ω of the $(n + 1)$-dimensional Euclidean space R^{n+1}. The following conditions are assumed valid throughout this chapter:

1. $f(t, x)$ is continuous on Ω;
2. $f(t, x)$ satisfies a local Lipschitz condition, with respect to x, in Ω.

Condition 2 means that for any compact (closed and bounded) subset of Ω, say K, there exists a positive number L_K, such that

$$\|f(t, x) - f(t, y)\| \le L_K \|x - y\| \tag{3.2}$$

for any $(t, x), (t, y) \in K$. For instance, condition 2 holds for any $f(t, x)$ that is continuously differentiable with respect to x for $(t, x) \in \Omega$ (see remark 2 to Theorem 2.1).

Consider now an arbitrary point $(t_0, x^0) \in \Omega$. Inasmuch as Ω is open, there exist positive numbers a and b such that

$$D = \{(t, x);\ |t - t_0| \le a,\ \|x - x^0\| \le b\} \subset \Omega. \tag{3.3}$$

If we restrict our considerations to D, then we see that the conditions of Theorem 2.2 are satisfied. Consequently, there exists a unique solution of (3.1)—say, $x = x(t)$—defined on $|t - t_0| \le \delta = \min\{a, bM^{-1}\}$, where $M = \sup\|f(t, x)\|$, satisfying the condition $x(t_0) = x^0$.

Because $(t_0, x^0) \in \Omega$ is arbitrary, we can state that through any point of Ω passes an integral curve of the system (3.1). Moreover, it is unique if we take into account a small neighborhood of the "initial" value t_0. Now, consider a solution $x = x(t)$ of system (3.1) defined on the interval (t_1, t_2). This means that $(t, x(t)) \in \Omega$ for $t \in (t_1, t_2)$, and that $x(t)$ is differentiable at any $t \in (t_1, t_2)$ and satisfies (3.1). It follows at once, from (3.1), that $\dot{x}(t)$ is also continuous on (t_1, t_2).

Several problems arise with respect to the preceding solution. The first we shall formulate here is that of *global uniqueness*. Assume we have another solution $x = \tilde{x}(t)$, such that $\tilde{x}(t_0) = x(t_0)$ for a given $t_0 \in (t_1, t_2)$. It is understood that the interval of definition of the solution $x(t)$ has common points with the interval (t_1, t_2). The property of local uniqueness (see Theorem 2.2) ensures that $\tilde{x}(t) \equiv x(t)$ on a "small" interval around t_0. The problem now is to decide whether $\tilde{x}(t) \equiv x(t)$, on their common interval of definition. We shall see that the answer is affirmative.

Another problem we shall consider is that of *continuation* of the solution $x = x(t)$. If we can find another solution of (3.1)—say, $x = y(t)$—such that the interval of definition of $y(t)$ contains the interval (t_1, t_2) and $x(t) \equiv y(t)$ on (t_1, t_2), then $x = x(t)$ is called *a continuable solution*. The solution $y(t)$ is then a *continuation* of the solution $x(t)$. We shall speak often of continuability at the right or at the left, the meaning of these terms being obvious. If the solution $x = x(t)$ cannot be continued, then it is called *saturated*. The answer to the continuation problem states that any continuable solution of (3.1) can be continued until we find a saturated solution.

Finally, the last problem we want to formulate concerns *the behavior of a saturated solution at the ends of its interval of existence*. Roughly speaking, the answer to this problem states that "the ends" of the solution are on the boundary of the domain Ω or at infinity (the last situation may occur only for unbounded Ω).

§3.2. Global Uniqueness

Let us consider first the problem of global uniqueness.

Theorem 3.1

Assume that $x(t)$ and $\tilde{x}(t)$ are solutions of the system (3.1), *such that their intervals of definition have common points and $x(t_0) = \tilde{x}(t_0)$ at such a point. Then $x(t) = \tilde{x}(t)$ on the common interval of definition.*

Proof. Assume that (t_1, t_2) is the interval where both solutions are defined. Then $t_0 \in (t_1, t_2)$. It suffices to prove that $x(t) = \tilde{x}(t)$ for $t_0 \le t < t_2$. The case $t < t_0$ can be reduced to the previous one if we change the independent variable: $t = -\tau$. In other words, it suffices to prove the uniqueness at the right of t_0.

Consider now a number T, such that $t_0 < T < t_2$. It will be fixed in the first step of the proof, but we want to point out that it can be chosen arbitrarily close to t_2. Let $K \subset \Omega$ be a compact set such that $(t, x(t))$, $(t, \tilde{x}(t)) \in K$ for $t \in [t_0, T]$. The existence of the set K, with the preceding property, is the consequence of the fact that both sets $\{(t, x(t)); t \in [t_0, T]\}$ and $\{(t, \tilde{x}(t); t \in [t_0, T]\}$ are compact, which follows easily from the continuity of $x(t)$ and $\tilde{x}(t)$. We know that there exists $L_K > 0$, with the property (3.2).

Denote by x^0 the common value of the solutions $x(t)$ and $\tilde{x}(t)$ at $t = t_0$. For $t \in [t_0, T]$, we shall have

$$x(t) = x^0 + \int_{t_0}^{t} f(s, x(s))\, ds \tag{3.4}$$

$$\tilde{x}(t) = x^0 + \int_{t_0}^{t} f(s, \tilde{x}(s))\, ds \tag{3.5}$$

from which we get

$$\|x(t) - \tilde{x}(t)\| \le L_K \int_{t_0}^{t} \|x(s) - \tilde{x}(s)\|\, ds. \tag{3.6}$$

Inequality (3.6) gives

$$\|x(t) - \tilde{x}(t)\| < \varepsilon + L_K \int_{t_0}^{t} \|x(s) - \tilde{x}(s)\|\, ds \tag{3.7}$$

for any $\varepsilon > 0$ and $t \in [t_0, T]$.

Inequality (3.7) is of Gronwall type, and we obtain

$$\|x(t) - \tilde{x}(t)\| < \varepsilon \exp[L_K(t - t_0)], \qquad t \in [t_0, T]. \tag{3.8}$$

Inasmuch as ε is arbitrary, inequality (3.8) implies that $x(t) = \tilde{x}(t)$ on $[t_0, T]$. On the other hand, T can be chosen arbitrarily close to t_2, which proves that $x(t) \equiv \tilde{x}(t)$ on $[t_0, t_2)$. Theorem 3.1. is thus proved.

§3.3. Global Existence and the Behavior of Saturated Solutions

Before proving the main result concerning the continuation of solutions, we shall give a characterization for continuable solutions.

Lemma

Assume that $x = x(t)$ *is a solution of the system* (3.1), *defined on the interval* $[t_0, t_2)$. *Then it can be continued at the right of* t_2, *if and only if the set*

$$\Gamma = \{(t, x(t)); t \in [t_0, t_2)\}$$

satisfies the following conditions:

(a) *It is bounded*;

(b) *The distance from* Γ *to the boundary of* Ω *is positive.*

Proof. Let us remark that Γ is the graph of the solution on $[t_0, t_2)$. Condition (a) is automatically satisfied when Ω is bounded. We shall prove now the necessity of conditions (a) and (b). If $x = x(t)$ is continuable at the right of t_2, we can find a solution $x = y(t)$ of (3.1) such that it is defined on $[t_0, t_2')$, with $t_2' > t_2$, and $x(t) \equiv y(t)$ on $[t_0, t_2)$. Consequently, there exists the lim $x(t)$ as $t \to t_2-$, and we can assume that $x = x(t)$ is continuous on $[t_0, t_2]$. This fact implies that $\bar{\Gamma}$ is a compact set, belonging to Ω. Hence, Γ is bounded and its distance to the boundary of Ω is positive.

To prove the sufficiency of conditions (a) and (b), we shall construct a solution of (3.1) that continues the solution $x = x(t)$. First, we shall prove the existence of the limit

$$\lim_{t \to t_2-} x(t). \tag{3.9}$$

Inasmuch as Γ satisfies (a) and (b), it belongs to a compact set $K \subset \Omega$ (e.g., its closure $\bar{\Gamma}$). The function $f(t, x(t))$ is then bounded on $[t_0, t_2)$—say, $\|f(t, x(t))\| \leq M$, $t \in [t_0, t_2)$. But (3.1) gives

$$x(t) - x(\tau) = \int_\tau^t f(s, x(s))\, ds \tag{3.10}$$

from which we obtain

$$\|x(t) - x(\tau)\| \leq M\,|t - \tau|, \qquad t, \tau \in [t_0, t_2). \tag{3.11}$$

Inequality (3.11) shows that the Cauchy's criterion for the existence of the limit (3.9) is verified. Consequently, if we define $x(t_2) = \lim x(t)$ as $t \to t_2-$, $x = x(t)$ becomes a continuous function on $[t_0, t_2]$.

Moreover, it can be shown that $x(t)$ is differentiable at the left for $t = t_2$. Indeed, let us make $\tau \to t_2$ in (3.10). We obtain

$$x(t) - x(t_2) = \int_{t_2}^{t} f(s, x(s))\, ds \tag{3.12}$$

which leads to

$$\frac{x(t) - x(t_2)}{t - t_2} - f(t_2, x(t_2)) = \frac{1}{t - t_2} \int_{t_2}^{t} \{f(s, x(s)) - f(t_2, x(t_2))\}\, ds \tag{3.13}$$

for $t < t_2$. But $t \to t_2-$ implies $s \to t_2-$ and $f(s, x(s)) \to f(t_2, x(t_2))$. It follows now from (3.13) that $\dot{x}(t_2) = f(t_2, x(t_2))$, the derivative being considered at the left of t_2.

Inasmuch as $(t_2, x(t_2)) \in \Omega$, there exists a unique solution passing through this point and defined in a neighborhood of t_2. Let us denote by $x = y(t)$ this solution and suppose it is defined on the interval

$$[t_2 - \delta, t_2 + \delta],$$

with $\delta > 0$. We now define the function

$$\tilde{x}(t) = \begin{cases} x(t) & t \in [t_0, t_2] \\ y(t) & t \in (t_2, t_2 + \delta]. \end{cases} \tag{3.14}$$

Equation (3.14) is obviously a solution of (3.1) on $[t_0, t_2 + \delta] \supset [t_0, t_2)$. The only point that needs a discussion is $t = t_2$, but we have seen that the derivative at the left exists and equals $f(t_2, x(t_2))$. The same is true for the derivative at the right of t_2, and, consequently, $\tilde{x}(t)$ satisfies (3.1) everywhere on $[t_0, t_2 + \delta]$. Thus, conditions (a) and (b) ensure the continuation of the solution and the lemma is proved.

We are now able to prove the main result concerning the connection between continuable and saturated solutions.

Theorem 3.2

Let $x = x(t)$ *be a continuable solution of the system* (3.1). *Then there exists a unique saturated solution of the same system, continuing* $x(t)$.

Proof. The uniqueness of the saturated solution, containing $x(t)$, follows from Theorem 3.1.

We have to construct this solution. Before proceeding further, let us make a remark of geometric nature. By Ω_m we denote the subset of Ω consisting of those points (t, x), such that $t^2 + \|x\|^2 \leq m$ and

$$d[(t, x), \partial\Omega] \geq m^{-1},$$

where m is a positive integer and d denotes the distance from (t, x) to the boundary of Ω. It can be seen easily that $\Omega_m \neq \phi$, for sufficiently large m. Moreover, $\Omega_m \subset \Omega_{m+1}$ and $\bigcup \Omega_m = \Omega$.

It is of primary interest for us that each Ω_m is bounded and its distance to the boundary of Ω is positive. From the preceding lemma, it follows that a solution of (3.1), defined on $[t_0, T)$, is saturated at the right if and only if its graph meets each Ω_m, as t increases to T. Indeed, a continuable solution has its graph in a set Ω_m, for a convenient m. Consequently, it suffices to construct a solution that continues the solution $x(t)$ and such that the preceding property holds.

Assume $x = x(t)$ is given on the interval $[t_0, t_1)$. As previously shown, any continuable solution can be considered as being defined on the closed interval, that is, on $[t_0, t_1]$ in our case. Its graph

$$\Gamma = \{(t, x(t));\, t \in [t_0, t_1]\}$$

belongs to a certain Ω_m—say, Ω_{m_1}.

We shall now prove that there exists a continuation of $x(t)$, the graph of which meets Ω_m, with arbitrarily large m. Inasmuch as $\Omega_{m_1} \subset \Omega_{m_1+1}$, it follows that for any $(\tau, \xi) \in \Omega_{m_1}$, the set

$$D_{\tau, \xi} = \{(t, x);\, |t - \tau| \leq \alpha,\, \|x - \xi\| \leq \alpha\}$$

for sufficiently small α belongs to Ω_{m_1+1}. Let us denote $M = \sup\|f(t, x)\|$, $(t, x) \in \Omega_{m_1+1}$.

Theorem 2.2 states that through any point $(\tau, \xi) \in \Omega_{m_1}$ there passes a unique solution of the system (3.1), defined on $|t - \tau| \leq \delta = \min\{\alpha, \alpha M^{-1}\}$. The main feature here is the fact that the length of the interval of definition does not depend on (τ, ξ), it being the same for all $(\tau, \xi) \in \Omega_{m_1}$.

It is obvious that $x = x(t)$ can be continued such that its continuation, denoted also by $x(t)$, can be defined on $[t_0, t_1 + \delta]$. If the graph on this interval belongs again to Ω_{m_1}, we can make another continuation; this time we shall find a continuation of $x(t)$, defined on $[t_0, t_1 + 2\delta]$. Because t is bounded in Ω_{m_1}, it is clear that after a finite number of steps, the procedure described will give us a continuation of $x(t)$ defined on an interval $[t_0, t_1 + p\delta]$, such that its graph belongs to a certain Ω_{m_2}, $m_2 > m_1$. Inasmuch as $(t, x(t)) \in \Omega_{m_2}$ on $[t_0, t_1 + p\delta]$, we can apply the same procedure as before, and so on. We obtain in the manner previously described a solution continuing $x(t)$, the graph of which meets Ω_m for arbitrary $m \geq m_1$. It is clearly a saturated solution of the system (3.1).

This ends the proof of Theorem 3.2. It is a *global* existence theorem. The last problem that we shall discuss here concerns the behavior of a saturated solution at the "ends" of its interval of existence. We shall again restrict our considerations to the case of solutions saturated at the right, or, as they are frequently called, *saturated in the future.*

Theorem 3.3

Assume $x = x(t)$ *is a saturated solution in the future of the system* (3.1), *defined on* $[t_0, T)$. *Then, any limit point of the set* $\{(t, x(t)\}$, *as* $t \to T$, *is either the point at infinity or a boundary point of* Ω.

Proof. Let $\{t_m\}$ be a sequence such that $\{t_m\} \subset [t_0, T)$ and $t_m \to T$ as $m \to \infty$. Assume that there exists $\lim(t_m, x(t_m))$ as $m \to \infty$, in the proper or generalized sense.

If $T = +\infty$, then $(t_m, x(t_m)) \to \infty$ as $m \to \infty$. If $T < \infty$ but $\lim \|x(t_m)\| = \infty$ as $m \to \infty$, then $(t_m, x(t_m)) \to \infty$ as $m \to \infty$. Conversely, if $(t_m, x(t_m)) \to \infty$ as $m \to \infty$, then at least one of the sequences $\{t_n\}$ or $\{\|x(t_m)\|\}$ tends toward infinity as $m \to \infty$.

The only case we have to consider is that corresponding to $T < +\infty$ and $\{x(t_m)\}$ convergent to a point X. In such case, we want to prove that $(T, X) \in \partial\Omega$.

Inasmuch as $(t_m, x(t_m)) \in \Omega$ for any m, it follows that $(T, X) \in \Omega$ or $(T, X) \in \partial\Omega$. It is our aim to show that only the last case can occur. Assume, on the contrary, that $(T, X) \in \Omega$. Without loss of generality, we can assume that $\{t_m\}$ is an increasing sequence.

Consider now a compact set K, such that $(T, X) \in K \subset \Omega$, the point (T, X) being an interior point of K. For instance, we can take $K = \Omega_m$, with sufficiently large m. We can assume $(t_m, x(t_m)) \in K$, $m \geq 1$, taking into account that $(t_m, x(t_m)) \to (T, X)$. We claim that for infinitely many m, there exists τ_m such that

$$t_m < \tau_m < t_{m+1}, \qquad (\tau_m, x(\tau_m)) \in \partial K. \tag{3.15}$$

Otherwise, we should have $(t, x(t)) \in K$ for $T - \eta \leq t < T$, which implies that $x(t)$ is continuable at the right of T; its graph on $[T - \eta, T)$ belongs to K, and, consequently, it is bounded and at positive distance from $\partial\Omega$.

We agree that τ_m is the smallest number in (t_m, t_{m+1}) satisfying (3.15). It is easy to see that

$$\lim_{m \to \infty} (\tau_m, x(\tau_m)) = (T, X). \tag{3.16}$$

Indeed,

$$d[(t_m, x(t_m)), (\tau_m, x(\tau_m))] = (|t_m - \tau_m|^2 + \|x(t_m) - x(\tau_m)\|^2)^{1/2}$$

and, considering that

$$\|x(t_m) - x(\tau_m)\| \leq M(t_m - \tau_m)$$

where $M = \sup\|f(t, x)\|$, $(t, x) \in K$, we see that (3.16) holds true.

But ∂K is a closed set and (3.15) implies $(T, X) \in \partial K$. On the other hand, we assumed (T, X) to be an interior point of K. The contradiction proves that our assumption cannot be true. Consequently, $(T, X) \in \partial\Omega$, which ends the proof.

§3.4. Dependence of Solutions on Initial Values

Consider again the system (3.1) under the assumptions we made in §3.1. If $(t_0, x^0) \in \Omega$, then there exists a unique saturated solution passing through this point. Let us denote it by $x = x(t; t_0, x^0)$, to make clear the fact that it depends on the initial condition. As function of t, $x(t; t_0, x^0)$ is defined on an interval $(t_1, t_2) \ni t_0$ and is continuously differentiable.

Properties (3.17), (3.18), and (3.19) hold:

$$\frac{d}{dt} x(t; t_0, x^0) = f(t, x(t; t_0, x^0)), \qquad t \in (t_1, t_2) \tag{3.17}$$

$$x(t_0; t_0, x^0) = x^0. \tag{3.18}$$

If

$$\xi = x(\tau, t_0, x^0), \quad \text{then} \quad x(t; \tau, \xi) = x(t; t_0, x^0). \tag{3.19}$$

The meaning of relations (3.17) and (3.18) is obvious. For (3.19), we shall give more details. Inasmuch as $x(\tau; \tau, \xi) = \xi = x(\tau; t_0, x^0)$, it follows from Theorem 3.1 that $x(t; \tau, \xi) \equiv x(t; t_0, x^0)$ on the common interval of existence. But this interval is obviously (t_1, t_2), if we consider that $x(t; \tau, \xi)$ is the saturated solution through (τ, ξ).

Nothing in the preceding considerations is concerned with how $x(t; t_0, x^0)$ depends on (t_0, x^0). Of course, the answer to this question depends on how much we assume with respect to $f(t, x)$. We are not going to discuss properties other than *continuity*.

Theorem 3.4

The function $x(t; t_0, x^0)$ is continuous in (t_0, x^0), in the following sense: If $[t_0, T)$ is the maximal interval of existence in the future, corresponding to (t_0, x^0) fixed, then for any $\tilde{t}$ with $t_0 < \tilde{t} < T$ and $\varepsilon > 0$, there *exists* $\delta(\varepsilon, \tilde{t}) > 0$, *such that* $x(t; \tau, \xi)$ *is defined on* $[t_0, \tilde{t}]$ *and*

$$\|x(t; \tau, \xi) - x(t; t_0, x^0)\| < \varepsilon, \qquad t \in [t_0, \tilde{t}] \tag{3.20}$$

as soon as

$$|\tau - t_0| < \delta, \qquad \|\xi - x_0\| < \delta. \tag{3.21}$$

Proof. We shall begin with the remark that for (τ, ξ) sufficiently near to (t_0, x^0), $x(t; \tau, \xi)$ is defined on an interval that contains a small neighborhood $[t_0, t_0 + h]$, $h > 0$, at the right of t_0.

Indeed, we can assume $(t_0, x^0), (\tau, \xi) \in \Omega_{m_0}$, for sufficiently large m_0. As shown in the proof of Theorem 3.2, each solution $x(t; \tau, \xi)$, with $(\tau, \xi) \in \Omega_{m_0}$, is defined at least on the interval $|t - \tau| \le \eta$, where η is independent of (τ, ξ).

Consequently, we can assume $x(t; \tau, \xi)$ defined on $[t_0, t_0 + h]$, $h > 0$, for all (τ, ξ) in a neighborhood of (t_0, x^0). Moreover, we can assume that $(t, x(t; \tau, \xi)) \in \Omega_{m_0+1}$ on $[\tau - \eta, \tau + \eta] \supset [t_0, t_0 + h]$, for all (τ, ξ) in the same neighborhood of (t_0, x^0). This last assertion follows also from the arguments involved in the proof of Theorem 3.2.

Finally, because $(t, x(t; t_0, x^0)) \in \Omega_m$ on $[t_0, \tilde{t}]$, together with its neighborhood

$$t_0 \le t \le \tilde{t}, \qquad \|x - x(t; t_0, x^0)\| < \varepsilon \tag{3.22}$$

for sufficiently small ε and large m, we can assume without loss of generality that this property holds for $m = m_0$.

From (3.17) and (3.18) we obtain

$$x(t; \tau, \xi) = \xi + \int_\tau^t f(s, x(s; \tau, \xi))\, ds \tag{3.23}$$

$$x(t; t_0, x^0) = x^0 + \int_{t_0}^t f(s, x(s; t_0, x^0))\, ds \tag{3.24}$$

both equations being satisfied on $[t_0, t_0 + h]$. We can derive from (3.23) and (3.24) the following inequality:

$$\|x(t; \tau, \xi) - x(t; t_0, x^0)\| \le \|\xi - x^0\| + \left| \int_\tau^{t_0} \|f(s, x(s; \tau, \xi))\|\, ds \right|$$
$$+ \int_{t_0}^t \|f(s, x(s; \tau, \xi)) - f(s, x(s; t_0, x^0))\|\, ds. \tag{3.25}$$

To reduce inequality (3.25) to an integral inequality of Gronwall type on $[t_0, t_0 + h]$, we need further arguments. First, if we denote by $M = \sup \|f(t, x)\|$, $(t, x) \in \Omega_{m_0+1}$, then

$$\left| \int_\tau^{t_0} \|f(s, x(s; \tau, \xi))\|\, ds \right| \le M\, |\tau - t_0| \tag{3.26}$$

for all (τ, ξ) in the prescribed neighborhood of (t_0, x^0). Second, if L is the Lipschitz constant corresponding to $f(t, x)$ in Ω_{m_0+1}, then

$$\|f(s, x(s; \tau, \xi)) - f(s, x(s; t_0, x^0))\| \leq L\,\|x(s; \tau, \xi) - x(s; t_0, x^0)\|. \tag{3.27}$$

From (3.25), (3.26), and (3.27), we get

$$\|x(t; \tau, \xi) - x(t; t_0, x^0)\| \leq \|\xi - x^0\| + M\,|\tau - t_0| + L\int_{t_0}^{t} \|x(s; \tau, \xi) - x(s; t_0, x^0)\|\, ds \tag{3.28}$$

the inequality being satisfied on $[t_0, t_0 + h]$.

Actually, inequality (3.28) holds for all $t \geq t_0$, such that $x(t; \tau, \xi)$ is defined and $(t, x(t; \tau, \xi)) \in \Omega_{m_0+1}$. It follows from the preceding considerations that the set of $t \geq t_0$ for which it holds is not empty. Applying Gronwall's lemma to (3.28), we obtain

$$\|x(t; \tau, \xi) - x(t; t_0, x^0)\| \leq (\|\xi - x^0\| + M\,|\tau - t_0|)\exp[L(t - t_0)] \tag{3.29}$$

for all $t \geq t_0$ such that (3.28) holds.

It is now our aim to prove that $x(t; \tau, \xi)$ is defined on $[t_0, \tilde{t}]$ and satisfies (3.20), if

$$(\|\xi - x^0\| + M\,|\tau - t_0|)\exp[L(\tilde{t} - t_0)] < \varepsilon \tag{3.30}$$

and (τ, ξ) belongs to the neighborhood of (t_0, x^0), described previously. Indeed, from (3.29) and (3.30) it follows that (3.20) is verified on $[t_0, t_0 + h]$. On the other hand, $x(t; \tau, \xi)$ exists for values of $t > t_0 + h$, because it is a saturated solution and its graph on $[t_0, t_0 + h]$ satisfies the conditions assuring the fact that it can be continued in the future. Condition (3.30) implies that $x(t; \tau, \xi)$ will satisfy (3.20) for all $t \in [t_0, \tilde{t}]$ for which it exists.

Assume, on the contrary, that the maximal interval of existence in the future for $x(t; \tau, \xi)$ is $[t_0, t_1)$ with $t_1 \leq \tilde{t}$. Inasmuch as the graph of a saturated solution leaves each Ω_m as t approaches the ends of the maximal interval of existence, our assumption would imply that there exists t_2, $t_0 + h < t_2 < t_1$, such that $\|x(t_2; \tau, \xi) - x(t_2; t_0, x^0)\| = \varepsilon$; i.e., the graph of $x(t; \tau, \xi)$ that lies in the set (3.22) for $t \in [t_0, t_0 + h]$, will meet the boundary of this set at $t = t_2$. But the preceding relation contradicts (3.29) and (3.30). Consequently, our assumption is not acceptable—that is, $t_1 > \tilde{t}$ and (3.20) holds. Theorem 3.4 is now completely proved.

Remark 1. Inequality (3.29) shows that $x(t; \tau, \xi)$ satisfies a local Lipschitz condition with respect to (τ, ξ). This property is, of course, stronger than that of continuity.

Inasmuch as $x(t; \tau, \xi)$ is continuously differentiable in t, it follows that it is continuous in $(t; \tau, \xi)$.

Remark 2. One can prove the differentiability of $x(t; \tau, \xi)$ with respect to (τ, ξ). It suffices to assume that $f(t, x)$ is differentiable in x (which implies, of course, the local Lipschitz condition).

§3.5. Differential Inequalities and the Comparison Method

We shall study now the scalar differential inequality

$$\dot{x}(t) \leq \omega(t, x(t)) \tag{3.31}$$

where $\omega(t, x)$ is a continuous function in the domain $\Delta \subset R^2$, satisfying in Δ the local Lipschitz condition. It is understood that $x(t)$ is a differentiable function on $[t_0, T)$, such that its graph on this interval belongs to Δ. Assume $x_0 \geq x(t_0)$ and consider the solution $x = \tilde{x}(t)$ of the equation

$$\dot{x} = \omega(t, x) \tag{3.32}$$

with initial condition $\tilde{x}(t_0) = x_0$. In other words, $\tilde{x}(t) = x(t; t_0, x_0)$. We assume also that $\tilde{x}(t)$ is saturated in the future.

The main result we want to establish can be formulated as follows.

Theorem 3.5

If $x(t)$ is a differentiable solution of (3.31) *and $\tilde{x}(t)$ is the solution of* (3.32) *with initial condition $\tilde{x}(t_0) = x_0 \geq x(t_0)$, then*

$$x(t) \leq \tilde{x}(t) \tag{3.33}$$

for all $t \geq t_0$, such that both sides in (3.33) *are defined.*

Proof. Assume, on the contrary, that inequality (3.33) does not hold. Then, there exists $t_1 > t_0$, such that

$$x(t_1) > \tilde{x}(t_1). \tag{3.34}$$

Consider now the function $y(t) = x(t) - \tilde{x}(t)$, and observe that $y(t_0) \leq 0$, $y(t_1) > 0$. Denote by τ the greatest value of t in $[t_0, t_1]$, such that $y(\tau) = 0$. The existence of τ is an immediate consequence of the continuity of $y(t)$. Indeed, the set $A = \{t; t \in [t_0, t_1], y(t) = 0\}$ is nonempty and closed, both properties being implied by the continuity of $y(t)$. The least upper bound of the set A will belong to A, and this least upper bound is nothing else but τ.

Inasmuch as we can find a compact set $K \subset \Delta$, such that $(t, x(t))$, $(t, \tilde{x}(t)) \in K$ for $t \in [\tau, t_1]$, we obtain

$$\dot{y}(t) = \dot{x}(t) - \dot{\tilde{x}}(t) \leq \omega(t, x(t)) - \omega(t, \tilde{x}(t)) \\ \leq L[x(t) - \tilde{x}(t)] = Ly(t).$$

Consequently, the function $y(t)$ satisfies the following conditions:

$$y(\tau) = 0, \qquad y(t) > 0, \qquad t \in (\tau, t_1], \tag{3.35}$$

$$\dot{y}(t) \leq Ly(t), \qquad t \in [\tau, t_1]. \tag{3.36}$$

We shall now prove that these conditions are not compatible. Indeed, from (3.36) we get (see §1.1)

$$y(t) \leq y(\tau) \exp[L(t - \tau)] = 0, \qquad t \in [\tau, t_1] \tag{3.37}$$

which contradicts (3.35). Theorem 3.5 is thus proved.

Remark. A similar result holds for differential inequalities of the form

$$\dot{x}(t) \geq \omega(t, x(t)). \tag{3.38}$$

It can be shown that $x(t) \geq \tilde{x}(t)$, where $\tilde{x}(t)$ is the solution of (3.32) such that $\tilde{x}(t_0) = x_0 \leq x(t_0)$.

We are now able to sketch the *comparison method* in the theory of ordinary differential equations. Roughly speaking, this method consists in reducing the study of some problems for differential systems to the study of corresponding problems for scalar differential equations.

More precisely, consider the system (3.1) with $f(t, x)$ continuous for $(t, x) \in \Omega$ and satisfying a local Lipschitz condition in the domain Ω. Let $V(t, x)$ be a real-valued function, differentiable on Ω. If $x = x(t)$ is a solution of (3.1), defined for $t \in (t_1, t_2)$, then $v(t) = V(t, x(t))$ is a differentiable function on (t_1, t_2), and

$$\dot{v}(t) = \frac{\partial V}{\partial t} + \left(\frac{\partial V}{\partial x}, f\right) \tag{3.39}$$

where $\partial V/\partial x$ denotes the vector having the components $(\partial V/\partial x_1), \ldots,$ $(\partial V/\partial x_n)$, and $(.,.)$ denotes, as usual, the scalar product:

$$\left(\frac{\partial V}{\partial x}, f\right) = \frac{\partial V}{\partial x_1} f_1 + \cdots + \frac{\partial V}{\partial x_n} f_n.$$

The right-hand side of (3.39) is called—when considered as function of (t, x)—*the derivative of* $V(t, x)$ *with respect to the system* (3.1). Assume now that $V(t, x)$ satisfies an inequality of the form

$$\frac{\partial V}{\partial t} + \left(\frac{\partial V}{\partial x}, f\right) \leq \omega(t, V), \qquad (t, x) \in \Omega. \tag{3.40}$$

This inequality is the one on which the comparison method leans. Concerning $\omega(t, y)$, we shall preserve the same assumptions we made previously. The domain Δ is generated by the point $(t, V(t, x))$, when $(t, x) \in \Omega$.

Let us consider now a solution of the system (3.1)—say, $x(t; t_0, x^0)$ for $t \in [t_0, T)$. If the real number y_0 is such that

$$V(t_0, x^0) \leq y_0 \tag{3.41}$$

then

$$V(t, x(t; t_0, x^0)) \leq y(t; t_0, y_0) \tag{3.42}$$

for all $t \geq t_0$, such that both sides are defined. Indeed, if we denote $v(t) = V(t, x(t; t_0, x^0))$, then (3.40) implies

$$\dot{v}(t) \leq \omega(t, v(t)), \qquad t \in [t_0, T). \tag{3.43}$$

Inequality (3.42) follows now from Theorem 3.5, where $y(t; t_0, y_0)$ is the solution of the equation

$$\dot{y} = \omega(t, y) \tag{3.44}$$

which is usually called the *equation of comparison*.

Of course, it is generally easier to get some information with respect to the scalar equation (3.44) than for the differential system (3.1). Once we have an estimate for $y(t; t_0, y_0)$, it is possible to obtain a corresponding one for $V(t, x(t; t_0, x^0))$ and—at last—some information about the solution $x(t; t_0, x^0)$.

The efficiency of the comparison method will be emphasized further in connection with the study of such global problems as the following: existence in the future of all solutions of a differential system of the form (3.1); existence of bounded solutions on a half-axis; the study of stability of solutions of ordinary differential systems.

§3.6. A Criterion of Global Existence

The purpose of this section is to apply the method of comparison to find conditions of global existence in the future for all the solutions of the system (3.1). We assume that $f(t, x)$ is a continuous vector function in the half-space $\Omega = \{(t, x); t \geq 0, x \in R^n\}$ and that a local Lipschitz condition holds in Ω.

Consider now a solution $x = x(t; t_0, x^0)$ of the system (3.1). From Theorem 3.2 it follows that this solution has a maximal interval of existence in the future—say, $[t_0, T)$, $T \leq \infty$. If $T = \infty$ for arbitrary $x^0 \in R^n$, we shall say that system (3.1) has the *property of global existence in the future*.

This statement means that any solution of (3.1) can be continued at the right, until we find a solution defined on the half-axis $t \geq t_0$.

It is necessary to point out that in the case $T < \infty$, the following property holds:

$$\lim_{t \to T-} \|x(t, t_0, x^0)\| = \infty. \tag{3.45}$$

Indeed, Theorem 3.3 states that any point (T, X), with $X = \lim x(t_n; t_0, x^0)$ as $t_m \to T$, lies on $\partial\Omega$ or is the point at infinity in R^{n+1}. Because Ω has no boundary points (T, X), with $T > 0$ and $X \in R^n$, (3.45) is established.

The general criterion of existence in the future, which we want to prove now, can be stated as follows.

Theorem 3.6 (Conti)

Consider the system (3.1) *with* $f(t, x)$ *continuous and satisfying a local Lipschitz condition in the half-space*

$$\Omega = \{(t, x), t \geq 0, x \in R^n\}. \tag{3.46}$$

Assume that there exists a differentiable function $V(t, x)$, *from* Ω *into* R, *satisfying the inequality* (3.40) *and such that*

$$V(t, x) \to +\infty \quad as \quad \|x\| \to +\infty \tag{3.47}$$

uniformly with respect to t in any compact set.

Then, if the comparison equation (3.44) *has the property of global existence in the future, the same property holds for the system* (3.1).

Proof. Let $t_0 \geq 0$ be given and consider the solution $x = x(t; t_0, x^0)$, with $x_0 \in R^n$, on its maximal interval of existence in the future—say, $[t_0, T)$.

Inasmuch as $V(t, x)$ satisfies (3.40), it follows that

$$V(t, x(t; t_0, x^0)) \leq y(t; t_0, V(t_0, x^0)) \tag{3.48}$$

for all $t \in [t_0, T)$. Indeed, (3.48) coincides with (3.42), where $y_0 = V(t_0, x^0)$. On the other hand, $y(t; t_0, y^0)$ is defined on $t \geq t_0$, for all real y_0.

Now, assume on the contrary that $T < \infty$. Then (3.45) holds for the solution $x(t; t_0, x^0)$, already chosen. From (3.48), making $t \to T$, we get a contradiction. Indeed, according to (3.45) and (3.47), the left side in (3.48) tends toward infinity as $t \to T$, whereas the right side tends toward $y(T; t_0, V(t_0, x^0))$—that is, is bounded above. This contradiction shows that our assumption $T < +\infty$ cannot be accepted. Consequently, $T = +\infty$, which means that system (3.1) has the property of global existence in the future.

Remark. It is clear from the proof of Theorem 3.6 that inequality (3.40) is necessary only for sufficiently large $\|x\|$. Indeed, instead of considering the solution on $[t_0, T)$, we can take into account only that part of the solution that corresponds to an interval $[t_1, T)$, with $t_1 > t_0$, such that $\|x(t; t_0, x^0)\|$ is large enough for $t \in [t_1, T)$.

Corollary

Consider the system (3.1) *in the half-space* $t \geq 0$, $x \in R^n$, *where* $f(t, x)$ *satisfies the preceding conditions and instead of* (3.40), *the following inequality*:

$$\|f(t, x)\| \leq \omega(t, \|x\|), \qquad t \geq 0, \qquad \|x\| > a. \tag{3.49}$$

Then, if $\omega(t, y)$ *is such that the comparison equation has the property of global existence in the future, the system* (3.1) *holds the same property.*

The proof is immediate, if we take in Theorem 3.6 $V(t, x) = \|x\|$, $\|x\| > a$. Then $V(t, x)$ is differentiable, and inasmuch as it does not contain t, we have

$$\frac{\partial V}{\partial t} + \left(\frac{\partial V}{\partial x}, f\right) = \frac{x_1 f_1 + x_2 f_2 + \cdots + x_n f_n}{(x_1{}^2 + x_2{}^2 + \cdots + x_n{}^2)^{1/2}}.$$

From Cauchy's inequality, we get

$$\frac{\partial V}{\partial t} + \left(\frac{\partial V}{\partial x}, f\right) \leq \frac{(x_1{}^2 + \cdots + x_n{}^2)^{1/2}(f_1{}^2 + \cdots + f_n{}^2)^{1/2}}{(x_1{}^2 + \cdots + x_n{}^2)^{1/2}} = \|f\|.$$

Taking (3.49) into account, we can write

$$\frac{\partial V}{\partial t} + \left(\frac{\partial V}{\partial x}, f\right) \leq \omega(t, V) \tag{3.50}$$

because $V(t, x) = \|x\|$. In other words, inequality (3.49) implies (3.40) for $V(t, x) = \|x\|$, $\|x\| > a$.

A noteworthy special case of the corollary is the so-called Wintner's criterion of global existence. It corresponds to the choice

$$\omega(t, y) = g(y) \tag{3.51}$$

with $g(y)$ such that

$$\int^{\infty} \frac{du}{g(u)} = +\infty. \tag{3.52}$$

Indeed, the comparison equation is

$$\dot{y} = g(y) \tag{3.53}$$

and from

$$\int_{y_0}^{y(t)} \frac{du}{g(u)} = t - t_0 \tag{3.54}$$

it follows that $y(t) \to \infty$ as $t \to T$ is impossible with $T < +\infty$ [see (3.52)]. Consequently, Eq. (3.53) has the property of global existence in the future, which proves the Wintner's criterion. We leave to the reader the task of formulating other special cases of Theorem 3.6 or of the corollary.

PROBLEMS

1. Let $A, B \subset R^n$ be two closed sets such that $d(A, B) = 0$. Does this property imply that $A \cap B \neq \varnothing$? Show that $d(A, B) > 0$ if A and B are disjoint closed sets and A is bounded (hence, compact).
2. Consider the linear equation $\dot{x} = a(t)x + b(t)$, with $a(t)$ and $b(t)$ continuous on (t_1, t_2). Find the function $x(t; t_0, x_0)$ and investigate its properties (e.g., the behavior at the endpoints of the interval of definition).
3. Find the function $x(t; t_0, x_0)$ in the case of the scalar equation $\dot{x} = -x^2$ and investigate its properties. In particular, find the maximal interval of existence of the solution passing through a given point. Illustrate the concept of saturated solution.
4. Consider the differential system S: $\dot{x} = f(t, x, \lambda)$, $x \in R^n$, where $f(t, x, \lambda)$ is a continuous vector function on $\Omega \times \Lambda$, with Ω a domain in the space R^{n+1} and Λ a neighborhood of $\lambda_0 \in R^m$. Assume further that $f(t, x, \lambda)$ satisfies a local Lipschitz condition with respect to the second argument. Denote by $x(t; t_0, x_0{}^0, \lambda)$ the solution of S passing through (t_0, x^0). If $x(t; t_0, x^0, \lambda_0)$ is defined for $t \in [a, b]$, prove that $x(t; \tau, \xi, \lambda)$ is defined on the same interval when (τ, ξ, λ) is sufficiently near to (t_0, x^0, λ_0). Investigate the continuity of the function $x(t; \tau, \xi, \lambda)$ by means of the method we used in the proof of Theorem 3.4.
5. Consider again the system S from problem 4 and let us assume, instead of a local Lipschitz condition with respect to x, that $f(t, x, \lambda)$ is continuously differentiable with respect to (x, λ). Show that $x(t; \tau, \xi, \lambda)$ is differentiable with respect to λ. The partial derivative of $x(t; \tau, \xi, \lambda)$ with respect to λ_j satisfies the linear differential system

   $$SV: \qquad \dot{y} = f_x[t, x(t; \tau, \xi, \lambda), \lambda]y + f_{\lambda_j}[t, x(t; \tau, \xi, \lambda), \lambda]$$

   and reduces to the null vector for $t = \tau$. The system SV is called the *variational system* associated to S and corresponding to the solution $x(t; \tau, \xi, \lambda)$. Here, f_x denotes the Jacobian matrix $(\partial f_i/\partial x_k)$, $i, k = 1, 2, \ldots, n$.
   Hint: Let e_j be the vector with m coordinates, such that all these coordinates are zero except that of rank j, which equals 1. For fixed τ and ξ, let us denote by $y(t; \lambda, h)$ the quantity

   $$x(t; \tau, \xi, \lambda + he_j) - x(t; \tau, \xi, \lambda) - hy(t)$$

   where $y(t)$ denotes the solution of SV such that $y(\tau) = 0$. Find for $y(t; \lambda, h)$ an integral inequality of Gronwall type.

6. From the preceding statement (problem 5), derive the following result concerning the differentiability of the solution with respect to the initial values. Consider the system $\dot{x} = f(t, x)$ under the assumption that its right side is continuous for $(t, x) \in \Omega \subset R^{n+1}$. Moreover, assume that the Jacobian matrix f_x is also continuous on the same set. Then $x(t; \tau, \xi)$ is differentiable with respect to ξ. The corresponding system in variations is

$$\dot{y} = f_x[t, x(t; \tau, \xi)]y.$$

Hint: The given differential system with the initial condition $x(\tau) = \xi$ can be transformed in $\dot{v} = f(t, v + \xi)$, with the initial condition $v(\tau) = 0$. Now ξ plays the role of the parameter λ from problem 5.

7. Consider the differential systems $\dot{x} = f(t, x)$ and $\dot{y} = f(t, y) + r(t, y)$, where $f(t, x)$ is continuous and locally Lipschitzian on $\Omega \subset R^{n+1}$ and $r(t, y)$ is continuous on the same set, and such that $\|r(t, y)\| \leq \varepsilon$. Let $x = x(t)$ be the solution of the first system, such that $x(t_0) = x^0$. Assume that $y = y(t)$ is a solution of the second system, defined at the right of t_0, and denote $y(t_0) = y^0$. Find an estimate for $\|y(t) - x(t)\|$.
Hint: It can be easily seen that $u(t) = \|y(t) - x(t)\|$ satisfies the following inequality of Gronwall type

$$u(t) \leq \|y^0 - x^0\| + \varepsilon(t - t_0) + L\int_{t_0}^{t} u(s)\, ds$$

where L is the Lipschitz constant for $f(t, x)$, corresponding to a compact set $K \subset \Omega$ such that K contains both graphs of $x(t)$ and $y(t)$ on the interval $[t_0, T]$, on which we need the estimate.

8. Consider the integral inequality

$$x(t) \leq M + \int_{t_0}^{t} \omega(s, x(s))\, ds$$

where $\omega(t, x)$ is a continuous function for $t \in [t_0, T)$, $x \in R$. Let us assume that a uniqueness condition holds for the comparison equation $\dot{y} = \omega(t, y)$ and that $\omega(t, x)$ is nondecreasing in x for each fixed t. Prove that $x(t) \leq y(t)$ on a certain interval $[t_0, T_1]$, $T_1 < T$, where $y(t)$ is the solution of the comparison equation such that $y(t_0) = M$.
Hint: Denote $z(t) = M + \int_{t_0}^{t} \omega(s, x(s))\, ds$ and notice that $x(t) \leq z(t)$. Find a differential inequality for $z(t)$.
Remark. This result generalizes Bihari's lemma.

9. Consider the differential system $\dot{x} = f(t, x)$ with $f(t, x)$ continuous and satisfying an inequality of the form $\|f(t, x) - f(t, y)\| \leq \omega(t, \|x - y\|)$. Find an estimate for the difference of two solutions, in terms of comparison equation $\dot{z} = \omega(t, z)$.

10. Consider the differential system $\dot{x} = f(t, x)$ and find the derivative of the function $V(t, x) = u(\|x\|)$ with respect to this system. Assume that $u(r)$ is differentiable, $x \neq 0$.

11. Consider the system $\dot{x} = f(t, x)$ and assume that $f(t, x)$ is continuous for $t \in [t_0, T)$ and $x \in R^n$. Let $f(t, x)$ be such that

$$\|f(t, x) - f(t, y)\| \leq \alpha(t)\|x - y\|, \ \|f(t, 0)\| \leq M\alpha(t) \exp\left[\int_{t_0}^{t} \alpha(s)\, ds\right].$$

Find an estimate for the solutions of the given system.

4

Some Special Classes of Differential Systems and Equations

The purpose of the present chapter is the study of two special classes of differential systems (or equations) that are frequently encountered in applications: namely, *linear systems* and *autonomous systems*. The former are characterized by the property of linearity of the right members with respect to the unknown functions, whereas the latter are such that the independent variable does not appear explicitly.

The last part of the chapter is devoted to the investigation of linear systems with constant coefficients. They are the only differential systems that are simultaneously linear and autonomous. We shall include a few concepts related to that of *exponential function of a matrix*, which are necessary to present in a concise form the theory of systems with constant coefficients.

§4.1. Linear Systems: Generalities

A linear first-order differential system can be written in the form

$$\dot{x}_i = \sum_{j=1}^{n} a_{ij}(t)x_j + b_i(t), \qquad i = 1, 2, \ldots, n \tag{4.1}$$

where $a_{ij}(t)$ and $b_i(t)$, $i, j = 1, 2, \ldots, n$ are defined on the interval (t_1, t_2). Usually, it is assumed that the functions $a_{ij}(t)$ and $b_i(t)$ are continuous on the preceding interval. If $b_i(t) = 0$, $i = 1, 2, \ldots, n$, then (4.1) becomes

$$\dot{x}_i = \sum_{j=1}^{n} a_{ij}(t)x_j, \qquad i = 1, 2, \ldots, n. \tag{4.2}$$

System (4.2) is called *homogeneous*, whereas (4.1) is generally termed *nonhomogeneous*.

It is convenient to introduce vector notation, so as to achieve more conciseness and elegance. Consider the column vector x with coordinates $x_1, x_2, \ldots, x_n$, the column vector $b(t)$ with coordinates $b_1(t), b_2(t), \ldots, b_n(t)$, and the square matrix $A(t)$ with elements $a_{ij}(t)$, $i, j = 1, 2, \ldots, n$. In full notation,

$$x = \begin{pmatrix} x_1 \\ x_2 \\ \vdots \\ x_n \end{pmatrix}, \qquad b(t) = \begin{pmatrix} b_1(t) \\ b_2(t) \\ \vdots \\ b_n(t) \end{pmatrix}, \qquad A(t) = \begin{pmatrix} a_{11}(t) & a_{12}(t) & \cdots & a_{1n}(t) \\ a_{21}(t) & a_{22}(t) & \cdots & a_{2n}(t) \\ \vdots & \vdots & \vdots & \vdots \\ a_{n1}(t) & a_{n2}(t) & \cdots & a_{nn}(t) \end{pmatrix}.$$

Then system (4.1) can be written

$$\dot{x} = A(t)x + b(t) \tag{4.3}$$

and the homogeneous system (4.2) becomes

$$\dot{x} = A(t)x. \tag{4.4}$$

Inasmuch as $A(t)$ is a matrix function, it is necessary to make precise some concepts we shall use further. The norm of a matrix $A = (a_{ij})$ with real elements is given by

$$\|A\| = \left(\sum_{i,j=1}^{n} a_{ij}^2 \right)^{1/2}. \tag{4.5}$$

It can be proved easily that $\|A + B\| \le \|A\| + \|B\|$, $\|\lambda A\| = |\lambda| \cdot \|A\|$ where λ is a scalar, $\|Ab\| \le \|A\| \cdot \|b\|$, where b is an n-vector.

By definition, the matrix function $A(t)$ is continuous if and only if all its elements $a_{ij}(t)$ are continuous functions. Using the norm, we can define the continuity in the usual way. Namely, $A(t)$ is continuous at t_0 if for any $\varepsilon > 0$, there exists $\delta(\varepsilon, t_0) > 0$ such that

$$\|A(t) - A(t_0)\| < \varepsilon \quad \text{for} \quad |t - t_0| < \delta.$$

The derivative or the integral of a matrix function can be defined in the standard way (i.e., the corresponding operation is applied to each element).

Let us consider now the existence problem for the linear system (4.3), with continuous $A(t)$ and $b(t)$ on (t_1, t_2). Assume that $t_0 \in (t_1, t_2)$ is fixed, and consider also an initial condition

$$x(t_0) = x^0, \qquad x^0 \in R^n. \tag{4.6}$$

Theorem 4.1

The solution $x(t; t_0, x^0)$ *of the system* (4.3) *is defined on* (t_1, t_2) *with respect to* t *and* t_0, *and for any* $x^0 \in R^n$.

Proof 1: Our system is of the form $\dot{x} = f(t, x)$, with $f(t, x) = A(t)x + b(t)$, where $A(t)$ and $b(t)$ are continuous on (t_1, t_2). Consequently, $f(t, x)$ is continuous on $\Omega = (t_1, t_2) \times R^n$. Inasmuch as

$$\|f(t, x) - f(t, y)\| \leq \|A(t)\| \cdot \|x - y\| \tag{4.7}$$

it follows that $f(t, x)$ satisfies a local Lipschitz condition with respect to x. More precisely, (4.7) shows that $f(t, x)$ satisfies a Lipschitz condition on each set $I \times R^n$, where $I \subset (t_1, t_2)$ is a compact interval. Indeed, $\sup\|A(t)\|$, for $t \in I$, is finite.

From the preceding considerations, it follows that $x(t; t_0, x^0)$ is uniquely determined for any (t_0, x^0), $t_0 \in (t_1, t_2)$, $x^0 \in R^n$. The interval of existence with respect to t, of the solution $x(t; t_0, x^0)$, contains a neighborhood of t_0 and belongs to (t_1, t_2). It is to be proved that this interval coincides with (t_1, t_2).

We shall prove now that $x(t; t_0, x^0)$ is defined on $[t_0, t_2)$. A similar argument holds in the case of the interval $(t_1, t_0]$. As usual, the last case is reduced to the former by means of the change of variable $t = -\tau$.

Assume that $[t_0, T)$, $T < t_2$, is the interval of existence in the future of the solution $x(t; t_0, x^0)$. Then

$$\lim\|x(t; t_0, x^0)\| = \infty, \quad \text{as} \quad t \to T. \tag{4.8}$$

But (4.3) implies

$$x(t; t_0, x^0) = x^0 + \int_{t_0}^{t} b(s)\, ds + \int_{t_0}^{t} A(s)x(s; t_0, x^0)\, ds \tag{4.9}$$

for $t \in [t_0, T)$. This yields

$$\|x(t; t_0, x^0)\| \leq M + \int_{t_0}^{t} \|A(s)\| \cdot \|x(s; t_0, x^0)\|\, ds \tag{4.10}$$

on the same interval, where M is such that

$$\|x^0\| + \int_{t_0}^{t} \|b(s)\|\, ds \leq M, \qquad t \in [t_0, T).$$

From (4.10) we get

$$\|x(t; t_0, x^0)\| \leq M \exp\left[\int_{t_0}^{t} \|A(s)\|\, ds\right], \qquad t \in [t_0, T) \tag{4.11}$$

which contradicts (4.8). Consequently, our assumption $T < t_2$ should be rejected—that is, we have $T = t_2$.

Proof 2: The method of successive approximations can be applied. We know that (4.3), (4.6) are equivalent to (4.9). For this last equation, the approximations are defined by

$$x^{m+1}(t; t_0, x^0) = x^0 + \int_{t_0}^{t} b(s)\, ds + \int_{t_0}^{t} A(s)x^m(s; t_0, x^0)\, ds \tag{4.12}$$

where $m = 0, 1, 2, \ldots$. Each $x^m(t; t_0, x^0)$ is obviously defined on (t_1, t_2).

To prove the convergence of the sequence $\{x^m(t; t_0, x^0)\}$, we conveniently restrict our considerations to a compact interval $I \subset (t_1, t_2)$. On such an interval, $\|A(t)\|$ is bounded, and we get from (4.12)

$$\|x^{m+1}(t; t_0, x^0) - x^m(t; t_0, x^0)\| \le L \left| \int_{t_0}^{t} \|x^m(s; t_0, x^0) - x^{m-1}(s; t_0, x^0)\|\, ds \right| \tag{4.13}$$

for any $m \ge 1$, where L is such that $\|A(t)\| \le L$, $t \in I$. Now, we can easily derive from (4.13) the well-known inequalities

$$\|x^{m+1}(t; t_0, x^0) - x^m(t; t_0, x^0)\| \le \frac{M}{L} \cdot \frac{(L|t - t_0|)^{m+1}}{(m+1)!} \tag{4.14}$$

where $M \ge L\|x^0\| + B$, with $B = \sup\|b(t)\|$, $t \in I$. In other words, the sequence $\{x^m(t; t_0, x^0)\}$ is uniformly convergent on any compact interval $I \subset (t_1, t_2)$.

Remark. Theorem 4.1 states that all the solutions of system (4.3) are defined on (t_1, t_2), if considered on their maximal interval of existence.

In our further developments concerning linear systems, we shall always understand by *solution* a *saturated solution*. Therefore, we shall handle only solutions defined on (t_1, t_2).

§4.2. Linear Homogeneous Systems

As mentioned, a linear homogeneous system has the form (4.4). The main property of linear homogeneous systems, from which many other properties can be derived, concerns the algebraic structure of the set of its solutions. It can be stated as follows.

Theorem 4.2

Consider the system (4.4) *in which x is an n vector and $A(t)$ is an n by n matrix, continuous on the interval* (t_1, t_2).

The set of all solutions of (4.4) *forms a vector space of dimension n, over the field of real numbers.*

Proof. Let us remark, from the beginning, that we are dealing with real solutions of real systems. If $A(t)$ is allowed to have complex-valued elements, or, if we consider complex solutions, we should also consider a vector space over the field of complex numbers.

That the set of all solutions forms a vector space can be seen from

$$(cx)^{\cdot} = c\dot{x} = cA(t)x = A(t)(cx) \tag{4.15}$$

and

$$(x + y)^{\cdot} = \dot{x} + \dot{y} = A(t)x + A(t)y = A(t)(x + y) \tag{4.16}$$

where x and y are solutions of (4.4) and c is an arbitrary real number. Relation (4.15) means that the product of a solution by a constant is also a solution, whereas (4.16) states that the sum of two solutions is also a solution. Of course, any linear (finite) combination of solutions is a solution [of system (4.4)].

To prove the fact that the space of solutions has dimension n, we shall construct a basis of this space consisting of n elements. Let us fix a $t_0 \in (t_1, t_2)$ and consider the following basis of the space R^n:

$$e^1 = \begin{pmatrix} 1 \\ 0 \\ 0 \\ \vdots \\ 0 \end{pmatrix}, \; e^2 = \begin{pmatrix} 0 \\ 1 \\ 0 \\ \vdots \\ 0 \end{pmatrix}, \; \ldots, \; e^n = \begin{pmatrix} 0 \\ 0 \\ 0 \\ \vdots \\ 1 \end{pmatrix}.$$

In other words, the vector e^k has all its coordinates zero, except the coordinate of rank k, which equals 1.

Denote by $x^k(t)$ the solution of (4.4) such that $x^k(t_0) = e^k$, $k = 1, 2, \ldots, n$. We shall prove that the solutions $x^k(t)$, $k = 1, 2, \ldots, m$, form a basis in the space of all solutions of system (4.4).

First, let us remark that these solutions are linearly independent. Indeed, from

$$c_1 x^1(t) + c_2 x^2(t) + \cdots + c_n x^n(t) = 0, \qquad t \in (t_1, t_2)$$

it follows—making $t = t_0$—that

$$c_1 e^1 + c_2 e^2 + \cdots + c_n e^n = 0$$

which implies $c_1 = c_2 = \cdots = c_n = 0$, because $e^1, e^2, \ldots, e^n$ form a basis in R^n.

Second, any solution $x(t)$ of (4.4) can be represented as a linear combination, with real coefficients, of the solutions $x^1(t)$, $x^2(t)$, $\ldots$, $x^n(t)$. Indeed, denote by $x(t)$ a solution of (4.4). We can write

$$x(t_0) = c_1 e^1 + c_2 e^2 + \cdots + c_n e^n \tag{4.17}$$

where $c_1, c_2, \ldots, c_n$ are some real numbers, uniquely determined. Denote by $y(t)$ the following solution of system (4.4):

$$y(t) = c_1 x^1(t) + c_2 x^2(t) + \cdots + c_n x^n(t) \tag{4.18}$$

the coefficients being those from (4.17). It is obvious that $x(t_0) = y(t_0)$, and inasmuch as the uniqueness holds, we can state that

$$x(t) = c_1 x^1(t) + c_2 x^2(t) + \cdots + c_n x^n(t), \qquad t \in (t_1, t_2) \tag{4.19}$$

which proves our assertion. Moreover, this representation is unique. Theorem 4.2 is thus proved.

Remark. From the preceding theorem, it follows that we know the set of all solutions of (4.4), as soon as we know n linearly independent solutions.

A basis of the space of all solutions of system (4.4) is called a *fundamental system* of solutions. Assume now we have n solutions of system (4.4)—say, $x^1(t), x^2(t), \ldots, x^n(t)$. Consider the matrix $X(t)$ the columns of which are the vectors $x^1(t), x^2(t), \ldots, x^n(t)$. We can write

$$X(t) = \begin{pmatrix} x_1{}^1(t) & x_1{}^2(t) & \cdots & x_1{}^n(t) \\ x_2{}^1(t) & x_2{}^2(t) & \cdots & x_2{}^n(t) \\ & & \vdots & \\ x_n{}^1(t) & x_n{}^2(t) & \cdots & x_n{}^n(t) \end{pmatrix} \tag{4.20}$$

the subscripts designing the coordinates and the superscripts the solutions.

Of course, an important question concerning system (4.4) is to decide whether $X(t)$ is the matrix of a fundamental system of solutions. The complete answer to this question is given in the next theorem.

Theorem 4.3

The necessary and sufficient condition that $x^1(t), x^2(t), \ldots, x^n(t)$ form a fundamental system for (4.4) *is*

$$\det X(t) \neq 0, \qquad t \in (t_1, t_2). \tag{4.21}$$

Proof. Assume $x^1(t), x^2(t), \ldots, x^n(t)$ form a fundamental system for (4.4) —that is, a basis in the space of all solutions of (4.4). For any solution $x(t)$ of system (4.4), there exists a unique system of constants $c_1, c_2, \ldots, c_n$, such that (4.19) holds. In other words, the linear algebraic system

$$c_1 x^1(t) + c_2 x^2(t) + \cdots + c_n x^n(t) = x(t) \tag{4.22}$$

has a unique solution (which does not depend on t) no matter how we choose the right member. This can occur only if (4.21) holds.

Conversely, if (4.21) holds, we can fix $t_0 \in (t_1, t_2)$ and represent any solution of (4.4) in the form (4.19). Indeed, because $\det X(t_0) \neq 0$, it follows that the vectors (in R^n) $x^1(t_0), x^2(t_0), \ldots, x^n(t_0)$ form a basis, and, for a given solution $x(t)$ of (4.4), there exist $c_1, c_2, \ldots, c_n$, such that

$$x(t_0) = c_1 x^1(t_0) + c_2 x^2(t_0) + \cdots + c_n x^n(t_0). \tag{4.23}$$

Following the same argument as in the proof of Theorem 4.2, we find that (4.19) holds with $c_1, c_2, \ldots, c_n$ from (4.22).

Remark. Inasmuch as in the proof of sufficiency of condition (4.21) we have used the weaker requirement $\det X(t_0) \neq 0$, for fixed $t_0 \in (t_1, t_2)$, it follows that (4.21) is equivalent also to this last condition.

In other words, if $\det X(t)$ is different from zero for a given $t_0 \in (t_1, t_2)$, it is different from zero everywhere on (t_1, t_2). An equivalent statement is that if $\det X(t)$ vanishes for a given $t_0 \in (t_1, t_2)$, then it vanishes identically on (t_1, t_2).

Let us notice that $\det X(t)$ is called the *Wronski determinant* of the system $x^1(t), x^2(t), \ldots, x^n(t)$. It is denoted sometimes as $W(x^1, x^2, \ldots, x^n)$, or, simply, $W(t)$, if confusion cannot occur.

We shall prove now a result concerning the Wronskian (Liouville's formula).

Theorem 4.4

Let $W(t)$ be the Wronskian of the solutions $x^1(t), x^2(t), \ldots, x^n(t)$ of (4.4). Then, if $t, t_0 \in (t_1, t_2)$, we have

$$W(t) = W(t_0) \exp\left[\int_{t_0}^{t} SpA(s)\, ds\right] \tag{4.24}$$

where

$$SpA(t) = \sum_{i=1}^{n} a_{ii}(t).$$

Proof. Inasmuch as

$$W(t) = \det X(t) = \begin{vmatrix} x_1{}^1(t) & x_1{}^2(t) & \cdots & x_1{}^n(t) \\ x_2{}^1(t) & x_2{}^2(t) & \cdots & x_2{}^n(t) \\ & & \vdots & \\ x_n{}^1(t) & x_n{}^2(t) & \cdots & x_n{}^n(t) \end{vmatrix}$$

we shall apply the rule of finding the derivative of a determinant. We can write

$$\dot{W}(t) = \sum_{k=1}^{n} W_k(t) \tag{4.25}$$

where $W_k(t)$ is the determinant with rows the same as in $W(t)$, except that of rank k, which consists in the derivatives $\dot{x}_k{}^1(t), \dot{x}_k{}^2(t), \ldots, \dot{x}_k{}^n(t)$. But

$$\dot{x}_k^i(t) = \sum_{j=1}^{n} a_{kj}(t) x_j^i(t), \qquad i, k = 1, 2, \ldots, n$$

which implies

$$W_k(t) = a_{kk}(t)W(t), \qquad k = 1, 2, \ldots, n. \tag{4.26}$$

Hence, from (4.25) and (4.26), we get

$$\dot{W}(t) = [SpA(t)]W(t). \tag{4.27}$$

Equation (4.27) yields (4.24), because it is a linear homogeneous equation in $W(t)$.

Remark. Formula (4.24) shows that $W(t)$ is either identically zero on (t_1, t_2), or everywhere different from zero.

We shall consider now the matrix differential equation associated with (4.4):

$$\dot{X} = A(t)X \tag{4.28}$$

We are mainly interested in finding n by n matrices satisfying (4.28). It follows from (4.28) that $X(t) = (x^1(t), x^2(t), \ldots, x^n(t))$—where $x^1(t), x^2(t), \ldots, x^n(t)$ are vector functions—is a solution of (4.28) if and only if each $x^k(t)$, $k = 1, 2, \ldots, n$, is a solution of (4.4). It is useful to notice that (4.28) has a unique solution $X(t)$, defined on (t_1, t_2), such that

$$X(t_0) = C \tag{4.29}$$

where $t_0 \in (t_1, t_2)$ and C is a constant n by n matrix.

Taking Theorem 4.3 into account, it follows that $X(t)$, defined by (4.28) and (4.29), is a *fundamental matrix* [i.e., the matrix of a fundamental system of solutions for (4.4)], if and only if $\det C \neq 0$. If we know a fundamental matrix for (4.4), or—which is the same—a solution of the matrix differential equation (4.28), such that $\det X(t) \neq 0$, then

$$x(t; t_0, x^0) = X(t)X^{-1}(t_0)x^0 \tag{4.30}$$

where $x(t; t_0, x^0)$ denotes the solution of (4.4) that takes the value x^0 for $t = t_0$. The inverse matrix in the right-hand side of (4.30) exists, because $\det X(t) \neq 0$, $t \in (t_1, t_2)$. We shall use further formula (4.30).

§4.3. Nonhomogeneous Linear Systems

Consider now the system (4.3)

$$\dot{x} = A(t)x + b(t)$$

where $A(t)$ is an n by n matrix function, continuous on (t_1, t_2), and where $b(t)$ is a continuous n vector function on the same interval. We shall assume we know a fundamental matrix $X(t)$ for the homogeneous system (4.4), and, applying the *method of variation of parameters*, we shall find the solution for (4.3).

Let us notice that formula (4.19), giving the general solution of (4.4), can be written as

$$x(t) = X(t)c \tag{4.31}$$

where c is the constant column vector with coordinates $c_1, c_2, \dots, c_n$. The method of variation of parameters consists in finding a solution of the form (4.31) for the nonhomogeneous system (4.3), with $c = c(t)$ conveniently chosen.

From (4.31), we find for $c = c(t)$

$$\dot{x}(t) = \dot{X}(t)c(t) + X(t)\dot{c}(t) = A(t)X(t)c(t) + X(t)\dot{c}(t)$$

and substituting in (4.3) we get

$$A(t)X(t)c(t) + X(t)\dot{c}(t) = A(t)X(t)c(t) + b(t)$$

from which we obtain

$$X(t)\dot{c}(t) = b(t). \tag{4.32}$$

Equation (4.32) defines uniquely $\dot{c}(t)$ on (t_1, t_2), inasmuch as $\det X(t) \neq 0$, $t \in (t_1, t_2)$. We obtain

$$\dot{c}(t) = X^{-1}(t)b(t) \tag{4.33}$$

which shows that

$$c(t) = c + \int_{t_0}^{t} X^{-1}(s)b(s)\,ds, \qquad t \in (t_1, t_2) \tag{4.34}$$

where c is a constant vector and $t_0 \in (t_1, t_2)$ is fixed. Formula (4.34) gives all $c(t)$ such that (4.33) holds.

Taking (4.34) into account, we find that (4.31) yields

$$x(t) = X(t)c + X(t)\int_{t_0}^{t} X^{-1}(s)b(s)\,ds, \qquad t \in (t_1, t_2) \tag{4.35}$$

with c an arbitrary constant vector. It is an elementary exercise to prove that (4.35) is, indeed, a solution of (4.3). Moreover, all the solutions of (4.3) are given by (4.35), with arbitrary constant c. To prove the last statement, it suffices to remark that for a convenient choice of c, $x(t)$ given by (4.35) satisfies the initial condition (4.6): $x(t_0) = x^0$. We get $x^0 = X(t_0)c$, which implies

$$x(t; t_0, x^0) = X(t)X^{-1}(t_0)x^0 + X(t)\int_{t_0}^{t} X^{-1}(s)b(s)\,ds \tag{4.36}$$

where $x(t; t_0, x^0)$ denotes the solution of (4.3), satisfying (4.6).

Sometimes, (4.36) is written as

$$x(t; t_0, x^0) = X(t)X^{-1}(t_0)x^0 + \int_{t_0}^{t} X(t)X^{-1}(s)b(s)\,ds, \tag{4.37}$$

inasmuch as

$$\tilde{x}(t) = X(t)\int_{t_0}^{t} X^{-1}(s)b(s)\,ds = \int_{t_0}^{t} X(t)X^{-1}(s)b(s)\,ds. \tag{4.38}$$

The last equality in (4.38) can be verified easily. It is also easy to see that

$$\dot{\tilde{x}}(t) = A(t)\tilde{x}(t) + b(t) \tag{4.39}$$

if we take into account Eq. (4.28) for $X(t)$. To obtain (4.39), it is convenient to use the first expression for $\tilde{x}(t)$ in (4.38).

Now, formula (4.35) is equivalent to

$$x(t) = X(t)c + \tilde{x}(t) \tag{4.40}$$

which can be read as follows: To obtain all the solutions of the nonhomogeneous system (4.3), it suffices to add the solution $\tilde{x}(t)$ of this system to an arbitrary solution of the homogeneous system (4.4). Actually, this statement holds true if $\tilde{x}(t)$ is replaced by another fixed solution of (4.3)—for instance,

$$x^*(t) = X(t)c^* + \tilde{x}(t).$$

Indeed, $X(t)c + x^*(t) = X(t)(c + c^*) + \tilde{x}(t) = X(t)c' + \tilde{x}(t)$, where c' is an arbitrary constant vector.

Consequently, as soon as we know the solutions of the homogeneous system (4.4)—that is, a fundamental matrix of this system—we can find a solution $\tilde{x}(t)$ of the nonhomogeneous system (4.3) by applying the method of variation of parameters. This operation requires, beside some algebraic calculations, the integration of a vector function, which means the integration of n scalar functions. Once we have found a solution $\tilde{x}(t)$ of (4.3), the general solution (or integral) of this system is given by (4.40).

§4.4. Linear Equations of Higher Order

We shall apply now the results we found in §4.2 and §4.3 to the case of linear differential equations of order n, with $n > 1$. Consider the nonhomogeneous linear equation

$$x^{(n)} = a_1(t)x^{(n-1)} + a_2(t)x^{(n-2)} + \cdots + a_n(t)x + b(t) \tag{4.41}$$

where $a_1(t), \ldots, a_n(t)$ and $b(t)$ are continuous scalar functions on (t_1, t_2).

We have seen in §2.3 that (4.41) is equivalent to

$$\begin{cases} \dot{x} = x_1 \\ \dot{x}_1 = x_2 \\ \quad\vdots \\ \dot{x}_{n-2} = x_{n-1} \\ \dot{x}_{n-1} = a_1(t)x_{n-1} + a_2(t)x_{n-2} + \cdots + a_n(t)x + b(t) \end{cases} \tag{4.42}$$

which represents a linear system of the form (4.3). Moreover, if $x(t)$ is a solution of (4.41), then

$$\begin{pmatrix} x(t) \\ \dot{x}(t) \\ \vdots \\ x^{(n-1)}(t) \end{pmatrix}$$

is a solution of (4.42). If we take an arbitrary solution of (4.42), then its first coordinate is a solution of (4.41).

Let us remark that the homogeneous equation corresponding to (4.41) is

$$x^{(n)} = a_1(t)x^{(n-1)} + a_2(t)x^{(n-2)} + \cdots + a_n(t)x \tag{4.43}$$

and the associated homogeneous system is

$$\begin{cases} \dot{x} = x_1 \\ \dot{x}_1 = x_2 \\ \quad\vdots \\ \dot{x}_{n-2} = x_{n-1} \\ \dot{x}_{n-1} = a_1(t)x_{n-1} + a_2(t)x_{n-2} + \cdots + a_n(t)x. \end{cases} \tag{4.44}$$

In view of the preceding correspondence between the solutions of (4.41) and (4.42)—respectively, (4.43) and (4.44)—we can derive several properties of linear higher order equations, starting from the properties of linear differential systems.

First, let us notice that from Theorem 4.1 we obtain a global result concerning the existence of solutions of linear higher order differential equations. More precisely, it follows that any solution of (4.41) is defined on (t_1, t_2). Theorem 4.2, applied to (4.44), gives the following property of the homogeneous linear equation (4.43): The set of all solutions of Eq. (4.43) forms a vector space of dimension n over the field of real numbers.

The linear dependence property of solutions of (4.43) can be formulated as follows: If $x_1(t), x_2(t), \ldots, x_m(t)$ are solutions of (4.43), we shall say that they are *linearly dependent* if there exist some constants $c_1, c_2, \ldots, c_m$, not all zero, such that

$$c_1x_1(t) + c_2 x_2(t) + \cdots + c_m x_m(t) = 0, \qquad t \in (t_1, t_2). \tag{4.45}$$

In the contrary case, they are called *linearly independent*.

Let us notice that from the corresponding concept of linear dependence, in the case of systems, we would get the following conditions instead of (4.45):

$$\begin{cases} c_1 x_1(t) + c_2 x_2(t) + \cdots + c_m x_m(t) = 0 \\ c_1 \dot{x}_1(t) + c_2 \dot{x}_2(t) + \cdots + c_m \dot{x}_m(t) = 0 \\ \vdots \\ c_1 x_1^{(n-1)}(t) + c_2 x_2^{(n-1)}(t) + \cdots + c_m x_m^{(n-1)}(t) = 0 \end{cases} \tag{4.46}$$

for arbitrary $t \in (t_1, t_2)$. Inasmuch as (4.46) implies (4.45), it suffices to prove the converse statement to establish their equivalence. Indeed, taking the derivatives of both sides in (4.45), we find (4.46). Consequently the linear dependence (or independence) concept has a simpler formulation in the case of linear equations than in the case of systems.

Of course, it is interesting to know whether a system of n solutions of (4.43) forms a fundamental system. Because the Wronskian corresponding to $x_1(t), x_2(t), \ldots, x_n(t)$ is

$$W(t) = \begin{vmatrix} x_1(t) & x_2(t) & \cdots & x_n(t) \\ \dot{x}_1(t) & \dot{x}_2(t) & \cdots & \dot{x}_n(t) \\ & & \vdots & \\ x_1^{(n-1)}(t) & x_2^{(n-1)}(t) & \cdots & x_n^{(n-1)}(t) \end{vmatrix} \tag{4.47}$$

the following statement results: A necessary and sufficient condition for a system of solutions $x_1(t), x_2(t), \ldots, x_n(t)$ of (4.43) to be a fundamental system is

$$W(t) \neq 0, \qquad t \in (t_1, t_2) \tag{4.48}$$

where $W(t)$ is given by (4.47). This statement is a special case of Theorem 4.3.

It is also useful to remark that Liouville's formula (4.24) becomes

$$W(t) = W(t_0) \exp\left[\int_{t_0}^{t} a_1(s)\, ds\right]. \tag{4.49}$$

Finally, we shall apply here the method of variation of parameters, to find a solution of Eq. (4.41), when a fundamental system of solutions is known for (4.43). System (4.32) becomes, in our case,

$$\begin{cases} \dot{c}_1 x_1(t) + \dot{c}_2 x_2(t) + \cdots + \dot{c}_n x_n(t) = 0 \\ \dot{c}_1 \dot{x}_1(t) + \dot{c}_2 \dot{x}_2(t) + \cdots + \dot{c}_n \dot{x}_n(t) = 0 \\ \vdots \\ \dot{c}_1 x_1^{(n-2)}(t) + \dot{c}_2 x_2^{(n-2)}(t) + \cdots + \dot{c}_n x_n^{(n-2)}(t) = 0 \\ \dot{c}_1 x_1^{(n-1)}(t) + \dot{c}_2 x_2^{(n-1)}(t) + \cdots + \dot{c}_n x_n^{(n-1)}(t) = b(t). \end{cases} \tag{4.50}$$

By hypothesis, the determinant of coefficients in (4.50) is different from zero [it is the Wronskian of the system $x_1(t)$, $x_2(t)$, $\ldots$, $x_n(t)$]. Consequently, $\dot{c}_1, \dot{c}_2, \ldots, \dot{c}_n$ are uniquely determined by (4.50), no matter how we take $b(t)$. Cramer's rule is applicable, and we find

$$\dot{c}_k(t) = \frac{W_k(t)}{W(t)}\, b(t), \qquad k = 1, 2, \ldots, n \tag{4.51}$$

where $W_k(t)$ is the determinant obtained from $W(t)$ by replacing the column of rank k by $0, 0, \ldots, 0, 1$.

Now, if we fix $t_0 \in (t_1, t_2)$, from (4.51) we get

$$c_k(t) = \int_{t_0}^{t} \frac{W_k(s)}{W(s)}\, b(s)\, ds, \qquad k = 1, 2, \ldots, n, \tag{4.52}$$

taking all integration constants as equal to zero. From (4.52), it follows that

$$\tilde{x}(t) = \sum_{k=1}^{n} x_k(t) \int_{t_0}^{t} \frac{W_k(s)}{W(s)}\, b(s)\, ds \tag{4.53}$$

is a solution of (4.41). Hence, the general integral of (4.41) is given by the formula

$$x(t) = c_1 x_1(t) + \cdots + c_n x_n(t) + \tilde{x}(t) \tag{4.54}$$

where $c_1, c_2, \ldots, c_n$ are arbitrary constants.

Remark. From (4.51), (4.52), and (4.53), it follows easily that $\tilde{x}(t)$ is that solution of (4.41) for which

$$\tilde{x}(t_0) = 0,\ \dot{\tilde{x}}(t_0) = 0, \ldots, \tilde{x}^{(n-1)}(t_0) = 0. \tag{4.55}$$

§4.5. Autonomous Systems

Another important class of differential systems, arising in many applications, is that of *autonomous systems*. Such a system can be written in the form

$$\dot{x}_i = f_i(x_1, x_2, \ldots, x_n), \qquad i = 1, 2, \ldots, n$$

or, using vector notation,

$$\dot{x} = f(x). \tag{4.56}$$

We shall assume that $f(x)$ is a vector function defined on a certain domain $D \subset R^n$, satisfying a local Lipschitz condition [of course, this implies the continuity of $f(x)$ in D].

From the general theorems established in Chapter 3, we can derive for the system (4.56) the following result: Given any $t_0 \in R$ and $x^0 \in D$, there exists a unique solution $x = x(t)$, defined on a maximal interval (t_1, t_2), such that $x(t_0) = x^0$.

Inasmuch as $f(x)$ does not depend on t, we can assume that it is a constant with respect to $t \in R$, and we take $\Omega = R \times D \subset R^{n+1}$ as the basic domain for developing the considerations of Chapter 3. On the other hand, taking into account these particularities, we feel the temptation to find some relationship between solutions corresponding to the same x^0 but different t_0's. This relationship will be made clear by the fact that

$$x(t; t_0, x^0) = x(t - t_0; 0, x^0) \tag{4.57}$$

both functions being defined on the same interval (t_1, t_2).

The meaning of (4.57) is the following: If we know the solutions of (4.56) corresponding to $t_0 = 0$ and arbitrary $x^0 \in D$, then we know all the solutions of this system. It suffices to apply all the translations along the t axis, to obtain all these solutions.

Before we proceed further with our investigation of autonomous systems, let us prove (4.57). Inasmuch as the right-hand side of (4.57) reduces to x^0 for $t = t_0$, we have to prove only that $x(t - t_0; 0, x^0)$ is a solution of (4.56). We know that

$$\dot{x}(t; 0, x^0) = f(x(t; 0, x^0))$$

for any t belonging to the maximal interval of existence of the solution passing by $(0, x^0)$—say, $(\tilde{t}_1, \tilde{t}_2)$. In view of the fact that f does not depend explicitly on t, it follows that

$$\dot{x}(t - t_0; 0, x^0) = f(x(t - t_0; 0, x^0))$$

for $t \in (\tilde{t}_1 + t_0, \tilde{t}_2 + t_0)$. Hence, both sides of (4.57) are (saturated) solutions of (4.56), taking the same value at $t = t_0$. They coincide, according to the global uniqueness theorem.

Let us denote

$$\tilde{x}(t; x^0) = x(t; 0, x^0). \tag{4.58}$$

The function $\tilde{x}(t; x^0)$ represents the solution of the system (4.56), taking the value x^0 at $t = 0$. It follows from the preceding considerations that $\tilde{x}(t; x^0)$ suffices to describe the set of all solutions of the system (4.56), because (4.57) can be written $x(t; t_0, x^0) = \tilde{x}(t - t_0; x^0)$.

We shall try now to take more advantage of the fact that our system is autonomous. First, it is interesting to point out that $x = \tilde{x}(t; x^0)$ is completely determined—as a curve in R^{n+1}—by its orthogonal projection on D

(i.e., on that section of Ω corresponding to $t = 0$), this projection being parametrized by means of the parameter t: $x = \tilde{x}(t; x^0)$. Hence, the projection of the point $(t, \tilde{x}(t; x^0)) \in R^{n+1}$ on D is $\tilde{x}(t; x^0)$. Consequently, there clearly appears the possibility of using the space R^n, instead of R^{n+1}, to obtain an adequate geometric interpretation of the behavior of integral curves of the system (4.56).

More precisely, let us consider in D all the curves of the equation $x = \tilde{x}(t; x^0)$, with $x^0 \in D$. Inasmuch as $x(t; t_0, x^0) = \tilde{x}(t - t_0; x^0)$, we can state that we have all the *trajectories* of system (4.56)—that is, the graphs of all solutions of this system. Of course, all the solutions $x = \tilde{x}(t - t_0; x^0)$, with $t_0 \in R$ and fixed x^0, have the same trajectory. As usual, we agree to consider the solutions on their maximal interval of existence.

Now, we shall prove that *two distinct trajectories cannot meet each other*. Indeed, assume we have $\tilde{x}(t_1; x^1) = \tilde{x}(t_2; x^2)$. Then $\tilde{x}(t; x^2) = \tilde{x}(t + t_1 - t_2; x^1)$, because both sides are solutions of (4.56) and they are equal for $t = t_2$, which shows that the trajectories under consideration are identical.

In view of a classification of the trajectories of an autonomous system, we need the following lemma.

Lemma

Let A be a closed subset of R that contains at least two elements, such that $a_1, a_2 \in A$ imply $a_1 + a_2 \in A$, $-a_1 \in A$. Then either $A = R$ or there exists $T > 0$ such that $A = \{nT\}$, where n is an arbitrary integer.

Proof. It is obvious that A contains at least one positive number. Now, we can distinguish only two distinct cases:

1. A contains a smallest positive number;
2. A contains arbitrarily small positive numbers.

Let us consider the first case and denote by T the smallest positive number belonging to A. It is then obvious that A contains any number of the form nT, with n an arbitrary integer. We shall prove that $A = \{nT\}$. Indeed, if $a \in A$, there exists n such that $nT \leq a < (n + 1)T$. But $a - nT \in A$, and from the preceding inequalities we get $0 \leq a - nT < T$. Inasmuch as T is the smallest positive number in A, it follows that $a - nT = 0$, that is, $a = nT$.

In the second case, there exists a sequence of positive numbers from A, such that $a_1 > a_2 > \cdots > a_n > \ldots$, $\lim a_n = 0$ as $n \to \infty$. Let a be an arbitrary real number. We shall prove that $a \in A$, —that is, $A = R$. Inasmuch as to any a_n there corresponds an integer k_n such that $k_n a_n \leq a < (k_n + 1)a_n$, it follows that $|a - k_n a_n| < a_n$. But $k_n a_n \in A$, and from

$\lim k_n a_n = a$ as $n \to \infty$, we obtain $a \in A$, the set A being a closed subset of R. Hence, $A = R$.

Remark. Another form of the lemma is as follows: The only proper closed subgroups of the additive group of real numbers are the discrete groups of the form $\{nT\}$, where T is a positive number.

We can prove now the main result concerning the classification of trajectories of an autonomous system.

Theorem 4.5

There are only three kinds of trajectories for an autonomous differential system:

1. *Trajectories that reduce to a point (singular point);*
2. *Closed trajectories, or cycles, that are homeomorphic to a circumference;*
3. *Open trajectories that are one-to-one continuous images of an open interval.*

Proof. Let $x = \tilde{x}(t; x^0)$ be the equation of a trajectory of the system (4.56), $x^0 \in D$ being fixed. If $\tilde{x}(t; x^0) \equiv x^0$, on the whole interval of existence of $\tilde{x}(t; x^0)$, then the first case occurs.

Assume now that $x(t; x^0) \not\equiv x^0$. There are only two possible situations. The first one corresponds to the case when the trajectory intersects itself, whereas the second characterizes the trajectories without autointersections [i.e., $\tilde{x}(t; x^0) \neq \tilde{x}(t_1; x^0)$ for $t \neq t_1$].

Consider the case of a trajectory that intersects itself. Then, if $x = \tilde{x}(t; x^0)$ is the equation of this trajectory, we can find two numbers t_1 and t_2, $t_1 \neq t_2$, both of them belonging to the maximal interval of definition of $\tilde{x}(t; x^0)$—say, $(\tilde{t}_1, \tilde{t}_2)$—such that $\tilde{x}(t_1; x^0) = \tilde{x}(t_2; x^0)$. We want to prove that $\tilde{t}_1 = -\infty$, $\tilde{t}_2 = +\infty$, and the trajectory is homeomorphic to a circumference. Indeed, $x = \tilde{x}(t + t_1 - t_2; x^0)$ is another solution of (4.56), corresponding to the same trajectory. But $\tilde{x}(t; x^0)$ and $\tilde{x}(t + t_1 - t_2; x^0)$ take the same value for $t = t_2$. The uniqueness theorem applies and we get

$$\tilde{x}(t; x^0) \equiv \tilde{x}(t + t_1 - t_2; x^0). \tag{4.59}$$

Inasmuch as both solutions are saturated, it follows that $(\tilde{t}_1, \tilde{t}_2)$ must coincide with $(\tilde{t}_1 + t_1 - t_2, \tilde{t}_2 + t_1 - t_2)$. This is possible if and only if $\tilde{t}_1 = -\infty$, $\tilde{t}_2 = +\infty$. On the other hand, (4.59) shows that $\tilde{x}(t; x^0)$ is a periodic function in t, with period $t_1 - t_2$. Because the set of all periods of a continuous function satisfies the conditions of the preceding lemma, it follows that a positive number T exists such that any period is of the form nT, with n an integer. In other words, T is the *fundamental period* of the solution $\tilde{x}(t; x^0)$. Let us point out that the case when any real number is a period of $\tilde{x}(t; x^0)$, would mean $\tilde{x}(t; x^0) \equiv x^0$, a situation we already considered.

Consequently, our trajectory corresponds to a periodic solution (of period T) of the system (4.56). To obtain all the trajectory, it suffices to restrict t to an arbitrary interval of length T—say, $[0, T)$. We have to prove that this trajectory is homeomorphic to a circumference. Inasmuch as any circumference is homeomorphic to the interval $[0, T]$, $T > 0$, with identified ends, we can consider the mapping $t \to \tilde{x}(t; x^0)$ from $[0, T]$ into R^n. This mapping is one-to-one, except $t = 0$ and $t = T$, for which $\tilde{x}(0; x^0) = \tilde{x}(T; x^0) = x^0$. Indeed, T is the smallest positive period of $\tilde{x}(t; x^0)$, which implies $\tilde{x}(t; x^0) \neq \tilde{x}(t_1; x^0)$ for $t \neq t_1$, $t, t_1 \in [0, T)$. Moreover, the mapping $t \to \tilde{x}(t; x^0)$ is continuous. Now, the following topological result[1] applies: A continuous one-to-one mapping of a compact space S into a Hausdorff space S_1 is homeomorphic (i.e., S is homeomorphic to its image in S_1). Thus, the needed conclusion is yielded.

Finally, in the case when the trajectory does not intersect itself, it follows that $t \to \tilde{x}(t; x^0)$ is a continuous one-to-one mapping from $(\tilde{t}_1, \tilde{t}_2)$—the maximal interval of existence—into R^n. From the preceding discussion, it follows that all possible situations are exhausted and Theorem 4.5 is thus proved.

Before concluding this section, we would like to make some comments concerning the terminology related to autonomous systems. It is usual to associate a vector field with the system (4.56). Namely, at each point $x \in D$ we consider the vector $f(x)$. Then the trajectories of the system (4.56) are nothing else but the curves that at each point are tangent to the vector $f(x)$. One usually refers to the space $R^n \supset D$ as the *phase space* of the dynamical system described by (4.56).

§4.6. Linear Systems and Equations with Constant Coefficients

A linear (homogeneous) system with constant coefficients has the form

$$\dot{x} = Ax \tag{4.60}$$

where x is a vector with n coordinates and A is an n by n matrix with constant elements. In scalar notation, (4.60) can be written

$$\dot{x}_i = \sum_{j=1}^{n} a_{ij} x_j, \qquad i = 1, 2, \ldots, n.$$

As previously stated, the linear systems with constant coefficients of the form (4.60) are the only linear (homogeneous) and autonomous differential

[1] For example, A. N. Kolmogorov and S. V. Fomin, *Elements of the Theory of Functions and Functional Analysis*, Graylock Press, Rochester, N. Y., 1957.

systems. Consequently, all results we found in the previous sections of this chapter apply to these systems.

Inasmuch as a fundamental matrix of (4.60) allows us to find all the solutions of this system, we shall construct such a matrix. There are several ways to approach this problem. It is our aim to present here two different solutions.

Consider the matrix equation associated with (4.60):

$$\dot{X} = AX. \tag{4.61}$$

We know (see §4.2) that any n by n matrix solution $X(t)$ of (4.61), such that $\det X(t) \neq 0$, is a fundamental matrix for the system (4.60). We know also that the condition $\det X(t) \neq 0$ is equivalent to $\det X(0) \neq 0$. To construct a fundamental matrix $X(t)$, the existence of which is guaranteed by Theorem 4.2, we shall define the matrix $\exp A$. This matrix is given by

$$\exp A = I + \frac{A}{1!} + \frac{A^2}{2!} + \cdots + \frac{A^m}{m!} + \cdots \tag{4.62}$$

where I is the unit n by n matrix.

Let us recall that the set of all n by n matrices is a vector space over the field of real numbers, and a norm $A \to \|A\|$ can be defined as shown in §4.1:

$$\|A\| = \left(\sum_{i,j=1} a_{ij}^2 \right)^{1/2}.$$

Moreover, this space is complete with respect to the norm. Indeed, it can be easily seen that a sequence of matrices—say, $\{A_p\}$—satisfying Cauchy's condition $\|A_p - A_q\| < \varepsilon$ as soon as $p, q \geq N(\varepsilon)$, for any $\varepsilon > 0$, is convergent to a matrix A. It suffices to notice that $\|A_p - A_q\| < \varepsilon$ implies $|a_{ij}{}^p - a_{ij}{}^q| < \varepsilon$, $i, j = 1, 2, \ldots, n$ and to apply the classical Cauchy's criterion.

Now, going back to (4.62), we denote

$$A_p = I + \frac{A}{1!} + \cdots + \frac{A^p}{p!}.$$

We shall prove that $\{A_p\}$ is a Cauchy sequence in the space of n by n matrices, as described. We obtain

$$A_{p+m} - A_p = \frac{A^{p+1}}{(p+1)!} + \cdots + \frac{A^{p+m}}{(p+m)!}$$

from which we find

$$\|A_{p+m} - A_p\| \leq \frac{\|A\|^{p+1}}{(p+1)!} + \cdots + \frac{\|A\|^{p+m}}{(p+m)!}. \tag{4.63}$$

Inasmuch as the series

$$\sum_{m=1}^{\infty} \frac{\|A\|^m}{m!}$$

is convergent, it follows from (4.63) that $\{A_p\}$ is a Cauchy sequence. This proves the existence of $\exp A$, defined by (4.62), for any n by n matrix A. One speaks also of the *exponential function of a matrix*.

Consider now

$$X(t) = \exp(At), \qquad t \in R. \tag{4.64}$$

We shall prove that $X(t)$ is the solution of (4.61), such that $X(0) = I$. Indeed, we have

$$\exp(At) = I + \frac{At}{1!} + \cdots + \frac{A^m t^m}{m!} + \cdots \tag{4.65}$$

which gives

$$\frac{d}{dt} \exp(At) = A + \frac{A^2 t}{1!} + \cdots + \frac{A^m t^{m-1}}{(m-1)!} + \cdots$$

that is,

$$\frac{d}{dt} \exp(At) = A \exp(At). \tag{4.66}$$

That $X(0) = I$ is seen from (4.65). Consequently, $X(t)$ given by (4.64) is a fundamental matrix for the system (4.60).

To get further information about the matrix $\exp(At)$, we need more properties of the space of matrices. At this point, it is convenient for us to assume that we are dealing with complex vectors and matrices (i.e., vectors the coordinates of which are complex numbers or matrices with complex elements). The preceding considerations remain valid with minor changes. For instance, in defining the norm of a complex vector, we shall take

$$\|x\| = (|x_1|^2 + \cdots + |x_n|^2)^{1/2}.$$

A similar change has to be made in the formula (4.5), giving $\|A\|$. The reason for considering complex vectors and matrices will appear soon.

A classical theorem of Jordan states that for any n by n complex matrix A there is a nonsingular matrix B, such that $B^{-1}AB = J$ has the canonical form

$$J = \begin{pmatrix} J_0 & 0 & 0 & \cdots & 0 \\ 0 & J_1 & 0 & \cdots & 0 \\ & & & \vdots & \\ 0 & 0 & 0 & \cdots & J_m \end{pmatrix} \tag{4.67}$$

where $J_0 = \operatorname{diag}(\lambda_1, \lambda_2, \ldots, \lambda_p)$ and

$$J_i = \begin{pmatrix} \lambda_{p+i} & 1 & 0 & \cdots & 0 & 0 \\ 0 & \lambda_{p+i} & 1 & \cdots & 0 & 0 \\ & & & \vdots & & \\ 0 & 0 & 0 & \cdots & \lambda_{p+i} & 1 \\ 0 & 0 & 0 & \cdots & 0 & \lambda_{p+i} \end{pmatrix} \tag{4.68}$$

for $i = 1, 2, \ldots, m$. Let p_i be the number of rows and columns of J_i (hence, $n = p + p_1 + \cdots + p_m$). The numbers $\lambda_j, j = 1, 2, \ldots, p + m$ are the *characteristic roots* of A. In general, they are complex even for real A.

Inasmuch as $A = BJB^{-1}$, it follows easily that $A^m = BJ^mB^{-1}$ for any m. Hence,

$$\exp(At) = B \exp(Jt)B^{-1}. \tag{4.69}$$

On the other hand, we obtain

$$\exp(Jt) = \begin{pmatrix} \exp(J_0 t) & 0 & \cdots & 0 \\ 0 & \exp(J_1 t) & \cdots & 0 \\ & & \vdots & \\ 0 & 0 & \cdots & \exp(J_m t) \end{pmatrix} \tag{4.70}$$

because the powers of the matrix J have the same structure as J itself. A simple calculation gives

$$\exp(J_0 t) = \operatorname{diag}[\exp(\lambda_1 t), \ldots, \exp(\lambda_p t)]. \tag{4.71}$$

It is somewhat more difficult to express $\exp(J_i t)$, $i = 1, 2, \ldots, m$. A few supplementary considerations are necessary. First, let us remark that J_i can be written in the form

$$J_i = \lambda_{p+i} I_{p_i} + Z_{p_i} \tag{4.72}$$

where I_{p_i} is the unit p_i by p_i matrix and Z_{p_i} denotes the p_i by p_i matrix of the form

$$Z_{p_i} = \begin{pmatrix} 0 & 1 & 0 & 0 & \cdots & 0 \\ 0 & 0 & 1 & 0 & \cdots & 0 \\ & & & & \vdots & \\ 0 & 0 & 0 & 0 & \cdots & 1 \\ 0 & 0 & 0 & 0 & \cdots & 0 \end{pmatrix}. \tag{4.73}$$

It is easy to obtain the powers of the matrix Z_{p_i}. For instance, to find $Z_{p_i}^2$ we have to move its diagonal, consisting of 1's, one place to the right from that of Z_{p_i}, and so on. It follows that $Z_{p_i}^{p_i}$ has all its elements zero.

Now, we need the formula

$$\exp(A + B) = (\exp A)(\exp B) \tag{4.74}$$

which is not true for arbitrary A and B. Nevertheless, it is valid for such pairs of matrices A and B that commute: $AB = BA$. This can be proven in the usual way, starting from (4.62). Inasmuch as I_{p_i} and Z_{p_i} commute, we shall get from (4.72) and (4.74)

$$\exp(J_i t) = \exp(\lambda_{p+i} I_{p_i} t) \cdot \exp(Z_{p_i} t). \tag{4.75}$$

From (4.62), it follows that

$$\exp(\lambda_{p+i} I_{p_i} t) = \exp(\lambda_{p+i} t) I_{p_i} \tag{4.76}$$

and

$$\exp(Z_{p_i} t) = \begin{pmatrix} 1 & t & \dfrac{t^2}{2!} & \cdots & \dfrac{t^{p_i - 1}}{(p_i - 1)!} \\ 0 & 1 & t & \cdots & \dfrac{t^{p_i - 2}}{(p_i - 2)!} \\ & & & \vdots & \\ 0 & 0 & 0 & \cdots & 1 \end{pmatrix}. \tag{4.77}$$

The last three formulas lead to the final expression for $\exp(J_i t)$:

$$\exp(J_i t) = \exp(\lambda_{p+i} t) \begin{pmatrix} 1 & t & \dfrac{t^2}{2!} & \cdots & \dfrac{t^{p_i - 1}}{(p_i - 1)!} \\ 0 & 1 & t & \cdots & \dfrac{t^{p_i - 2}}{(p_i - 2)!} \\ & & & \vdots & \\ 0 & 0 & 0 & \cdots & 1 \end{pmatrix}. \tag{4.78}$$

Consequently, if the canonical Jordan form J of the matrix A is known, then $\exp(At)$ is given by (4.69), with $\exp(Jt)$ expressed by (4.70) and (4.78). An immediate consequence of these formulas is that the elements of the matrix $\exp(At)$ are of the form

$$q(t) = \sum_{k=0}^{p+m} p_k(t) \exp(\lambda_k t) \tag{4.79}$$

where $p_k(t)$ are polynomials in t.

A function $q(t)$ of the form (4.79) will be called a *quasi-polynomial.* Therefore, we can state that for any linear homogeneous differential system, there exists a fundamental matrix the elements of which are quasi-polynomials. Moreover, the exponential functions appearing in these quasi-polynomials are of the form $\exp(\lambda t)$, with λ a characteristic root of the matrix of coefficients. We shall prove now a result concerning the quasi-polynomials, which is necessary in our further developments and presents much interest in the applications.

Lemma

Consider a quasi-polynomial

$$q(t) = \sum_{k=1}^{n} p_k(t) \exp(\lambda_k t) \tag{4.80}$$

with $\lambda_k \neq \lambda_j$ *for* $k \neq j$, *and assume* $q(t) \equiv 0$ *on the real axis. Then* $p_k(t) \equiv 0$, $k = 1, 2, \ldots, n$.

Proof. The statement is obvious for $n = 1$ because the exponential function does not vanish.

Assume now that the conclusion of the lemma is true for any quasi-polynomial with $(n - 1)$ exponential functions in its representation. Inasmuch as $q_1(t) = q(t) \exp(-\lambda_1 t)$ is itself a quasi-polynomial identically vanishing on the real axis, it follows that its derivatives of any order will be identically zero. If the degree of $p_1(t)$ is r, we shall take the derivative of order $r + 1$, $\tilde{q}(t) = q_1{}^{(r+1)}(t)$. Quasi-polynomial $\tilde{q}(t)$ is of the form

$$\tilde{q}(t) = \sum_{k=2}^{n} \tilde{p}_k(t) \exp[(\lambda_k - \lambda_1)t].$$

Because $\tilde{q}(t)$ contains only $(n - 1)$ exponential functions, our hypothesis of induction gives $\tilde{p}_k(t) \equiv 0$, $k = 2, 3, \ldots, n$.

We can prove that $\tilde{p}_k(t) \equiv 0$ is possible if and only if $p_k(t) \equiv 0$. Indeed, for fixed k we can write $p_k(t) = a_0 t^m + a_1 t^{m-1} + \cdots + a_m$, where $a_0 \neq 0$. A simple calculation shows that $\tilde{p}_k(t) = (\lambda_k - \lambda_1)^{r+1} a_0 t^m +$ terms of degree less than m. Hence, $p_k(t) \not\equiv 0$ implies $\tilde{p}_k(t) \not\equiv 0$, which ends the proof.

Consider again the system (4.60), and show the way to reduce it to an nth-order linear equation with constant coefficients. An algebraic auxiliary result will play an important role in this approach.

Assume that the characteristic equation of the matrix A, $\det(\lambda I - A) = 0$, has the form

$$\lambda^n + a_1 \lambda^{n-1} + \cdots + a_{n-1}\lambda + a_n = 0. \tag{4.81}$$

A theorem of Cayley[2] states that the matrix A is a "matrix solution" of its

[2] For example, A. Lichnerowicz, *Linear Algebra and Analysis*, Holden-Day, Inc., San Francisco, 1967.

own characteristic equation—that is,

$$A^n + a_1 A^{n-1} + \cdots + a_{n-1} A + a_n I = 0 \tag{4.82}$$

where 0 is the matrix the elements of which are all zero. From (4.60), we get $\ddot{x} = A\dot{x} = A(Ax) = A^2 x$, $\dddot{x} = A^3 x$, and so on.

On the other hand, for any vector x with n coordinates, we find, from (4.82),

$$A^n x + a_1 A^{n-1} x + \cdots + a_{n-1} A x + a_n x = 0. \tag{4.83}$$

Consequently, for any solution x of (4.60), we obtain

$$x^{(n)} + a_1 x^{(n-1)} + \cdots + a_{n-1} \dot{x} + a_n x = 0. \tag{4.84}$$

That is, the vector x is a solution of an nth-order differential equation with constant coefficients. More precisely, each coordinate of the vector solution x of the system (4.60) satisfies the same linear scalar differential equation of order n, with constant coefficients.

In what follows, we shall investigate Eq. (4.84), under the assumption that x is a scalar function. As already pointed out, this suffices also for the solution of linear systems.

Let us denote

$$p(\lambda) = \lambda^n + a_1 \lambda^{n-1} + \cdots + a_{n-1} \lambda + a_n \tag{4.85}$$

and

$$L(x) = x^{(n)} + a_1 x^{(n-1)} + \cdots + a_{n-1} \dot{x} + a_n x. \tag{4.86}$$

The polynomial $p(\lambda)$ is called the *characteristic polynomial* associated to (4.84), and, accordingly, the equation $p(\lambda) = 0$ is called the *characteristic equation*. Of course, $L(x)$ has a meaning only for those x that are differentiable at least of order n. Using this notation, we can write Eq. (4.84) as $L(x) = 0$. It is interesting to notice that

$$L(e^{\lambda t}) = e^{\lambda t} p(\lambda). \tag{4.87}$$

As a consequence of (4.87), we can mention the following fact: The exponential function $e^{\lambda t}$ is a solution of (4.84) if and only if λ is a root of the characteristic equation. Hence, if the equation $p(\lambda) = 0$ has n distinct roots—say, $\lambda_1, \ldots, \lambda_n$, we can expect that

$$\exp(\lambda_1 t), \exp(\lambda_2 t), \ldots, \exp(\lambda_n t)$$

is a fundamental system of solutions for (4.84). Indeed, their Wronskian is not equal to zero, as can be easily seen.

Assume now that the equation $p(\lambda) = 0$ has $m < n$ distinct roots, $\lambda_1, \lambda_2, \ldots, \lambda_m$, their multiplicities being $k_1, k_2, \ldots, k_m$. A fundamental system of solutions can be constructed, as shown in the next theorem.

Theorem 4.6

With the preceding notations, the functions $t^k \exp(\lambda_j t)$, $k = 0, 1, \ldots, k_j - 1$, $j = 1, 2, \ldots, m$, *form a fundamental system of solutions for* (4.84).

Proof. The basic property we need in this proof is expressed by the following formula:

$$L(t^k e^{\lambda t}) = [p^{(k)}(\lambda) + kp^{(k-1)}(\lambda)t + \frac{k(k-1)}{2!} p^{(k-2)}(\lambda)t^2 + \cdots + p(\lambda)t^k]e^{\lambda t}. \tag{4.88}$$

The proof of (4.88) can be obtained immediately. Indeed, $t^k e^{\lambda t} = (d^k/d\lambda^k)(e^{\lambda t})$, which implies

$$L(t^k e^{\lambda t}) = \frac{d^k}{d\lambda^k} L(e^{\lambda t}) = \frac{d^k}{d\lambda^k} [p(\lambda)e^{\lambda t}]$$

taking (4.87) into account. Formula (4.88) is now obvious.

Assume that $\lambda = \lambda_j$. Then $p(\lambda) = p'(\lambda) = \cdots = p^{(k_j - 1)}(\lambda) = 0$. From (4.88) we derive $L(t^k e^{\lambda t}) = 0$, $k = 0, 1, \ldots, k_j - 1$. Consequently, we constructed $n = k_1 + \cdots + k_n$ solutions for (4.84). It remains to prove that they are linearly independent. Indeed, from

$$\sum_{j=1}^{m} (c_{j,1} + c_{j,2}t + \cdots + c_{j,k_j-1}t^{k_j-1}) \exp(\lambda_j t) \equiv 0$$

it follows, by applying the lemma, that all the constants $c_{j,k}$ are zero, which proves the theorem.

Before concluding this section, we shall make a few comments concerning the case of systems (or equations) with real coefficients. For example, assume that Eq. (4.84) has real coefficients. The construction given previously for the fundamental system of solutions does not exclude the possibility of complex solutions. The main question is, how can we construct a fundamental system consisting of real solutions?

The answer is quite simple. If λ_j is a real root of the characteristic equation, then $t^k \exp(\lambda_j t)$ are real solutions. When λ_j is complex, there is another characteristic root—say, λ_{j+1}—such that $\lambda_{j+1} = \bar{\lambda}_j$. They have the same multiplicity, and starting from the set of $2k_j$ complex solutions $\{t^k \exp(\lambda_j t), t^k \exp(\bar{\lambda}_j t)\}$, $k = 0, 1, \ldots, k_j - 1$, we can form the following set of $2k_j$ real solutions: $\{t^k \exp(\mu_j t) \cos \nu_j t, t^k \exp(\mu_j t) \sin \nu_j t\}$, $k = 0, 1,$

$2, \ldots, k_j - 1$, with $\mu_j = \operatorname{Re} \lambda_j$, $\nu_j = \operatorname{Im} \lambda_j$. This procedure applies with minor modifications in the case of linear systems with constant coefficients.

As can be seen from the preceding considerations, the main fact we have to take into account is that the real and imaginary parts of a complex solution of a system (equation) with real coefficients are also solutions of the same system (equation).

PROBLEMS

1. Consider the differential system $S: \dot{x} = A(t)x$, with $A(t)$ continuous on (t_1, t_2). Let $T(t)$ be a nonsingular matrix, continuously differentiable on (t_1, t_2). Then $y = Tx$ satisfies the linear system, $\bar{S}$: $\dot{y} = (TAT^{-1} + \dot{T}T^{-1})y$. If $A(t)$ is a constant matrix, find $T =$ const, such that $\bar{S}$ is an upper triangular system (i.e., the matrix of its coefficients has only zero elements under the principal diagonal).
2. Let $S: \dot{x} = A(t)x$ be a system with $A(t)$ continuous and bounded on the half-axis $t \geq 0$. Prove that any solution $x = x(t) \neq 0$ of S satisfies
$$\limsup_{t\to\infty} \frac{\ln \|x(t)\|}{t} < +\infty.$$
Hint: Use the Gronwall's inequality.
3. Consider the linear system $S: \dot{x} = A(t)x$, where $A(t)$ is a continuous real matrix on (t_1, t_2). Let $A^*(t)$ be the transpose of $A(t)$, and consider the *adjoint system* S^*: $\dot{y} = -A^*(t)y$. Prove that the scalar product of a solution of S by an arbitrary solution of S^* is a constant on (t_1, t_2).
4. (Ważewski's inequality.) With the notations from the preceding problem, let us define $H(t) = [A(t) + A^*(t)]/2$. Inasmuch as $H(t)$ has only real characteristic roots, we can consider the smallest root $\lambda(t)$ as well as the largest characteristic root of $H(t)$—say, $\Lambda(t)$. If $t_0 \in (t_1, t_2)$ and $t \geq t_0$, then any solution $x = x(t)$ of the system S satisfies
$$\|x(t_0)\| \exp\left[\int_{t_0}^{t} \lambda(s)\, ds\right] \leq \|x(t)\| \leq \|x(t_0)\| \exp\left[\int_{t_0}^{t} \Lambda(s)\, ds\right].$$
Hint: Denote $y(t) = (x(t), x(t)) = \|x(t)\|^2$, and show first that $\dot{y} = 2(x, Hx)$. Inasmuch as we have $\lambda(t)y \leq (x, Hx) \leq \Lambda(t)y$, the inequality can be obtained by integration.
5. Consider the homogeneous linear equation (4.43), and assume we know a solution $u(t) \neq 0$. Prove that $y = x/u$ satisfies an equation of the same form, the coefficient of y being identically zero. Then reduce the equation to an equation of order $(n - 1)$.
6. Consider the differential equation $\ddot{x} + \omega^2 x = f(t)$, with $f(t)$ continuous on (t_1, t_2), ω being a real constant. Show that the general integral is given by
$$x(t) = A \cos \omega t + B \sin \omega t + \frac{1}{\omega} \int_{t_0}^{t} \sin \omega(t - s) f(s)\, ds,$$
where A and B are arbitrary constants and t_0 is a fixed number in (t_1, t_2). Use the preceding formula to find an integral equation that is equivalent to the nonlinear differential equation $\ddot{x} + \omega^2 x = f(t, x)$.

7. Show that the second-order linear differential equation $a_0(t)\ddot{x} + a_1(t)\dot{x} + a_2(t)x = 0$, where $a_i(t)$, $i = 0, 1, 2$ are continuous on the interval (t_1, t_2) and $a_0(t) \neq 0$, can be written in the form

$$\frac{d}{dt}\left[p(t)\frac{dx}{dt}\right] + q(t)x = 0, \qquad p(t) \neq 0.$$

Find $p(t)$ and $q(t)$.
Hint: Find $\eta = \eta(t)$ such that the equation obtained from the given one by multiplying with $\eta(t)$ has the form $p(t)\ddot{x} + \dot{p}(t)\dot{x} + q(t)x = 0$.

8. Consider the Bessel differential equation $t^2\ddot{x} + t\dot{x} + (t^2 - n^2)x = 0$, where n is a nonnegative integer. Using the method of undetermined coefficients, find solutions of the form

$$x(t) = t^{\alpha} \sum_{k=0}^{\infty} a_k t^k, \qquad a_0 \neq 0.$$

Show that

$$x(t) = \sum_{k=0}^{\infty} \frac{(-1)^k (x/2)^{n+2k}}{k!(n+k)!}$$

is a solution.

9. Using the method of undetermined coefficients, find series solutions of the Gauss hypergeometric equation

$$t(1-t)\ddot{x} + [c - (a+b+1)t]\dot{x} - abx = 0.$$

In particular, show that

$$x(t) = 1 + \frac{ab}{1!\,c}t + \frac{a(a+1)b(b+1)}{2!\,c(c+1)}t^2 + \cdots + \frac{a(a+1)\cdots(a+n-1)b(b+1)\cdots(b+n-1)}{n!\,c(c+1)\cdots(c+n-1)}t^n + \cdots$$

represents a solution (the hypergeometric series).

10. Consider the Legendre's equation $(1-t^2)\ddot{x} - 2t\dot{x} + \lambda x = 0$, where λ is a parameter. Show that there exist polynomial solutions if and only if $\lambda = n(n+1)$.

11. Let $\dot{x} = f(x)$ be an autonom ous system, with $f(x)$ satisfying a local Lipschitz condition in R^n. Consider a solution $x = x(t)$, which exists for $t \geq 0$. Let $\Omega[x(t)]$ be the set of all points in R^n such that there exist sequences $\{t_n\}$, $t_n \to \infty$, with $x(t_n) \to \bar{x} \in \Omega$ as $n \to \infty$. Prove that Ω is a closed set consisting of trajectories of the given system. If $x(t) \to \bar{x}$ as $t \to \infty$, then $\bar{x}$ is a singular point of the system.
Hint: If $x(t) = \tilde{x}(t; x^0)$, then $\tilde{x}(t; \tilde{x}(s; x^0)) = \tilde{x}(t+s; x^0)$, $t, s \geq 0$. Let $\bar{x} = \lim_{n\to\infty} \tilde{x}(t_n; x^0)$ be an arbitrary point in Ω. Then $\tilde{x}(t; \bar{x}) \in \Omega$ because $\tilde{x}(t+t_n; x^0) = \tilde{x}(t; \tilde{x}(t_n; x^0)) \to \tilde{x}(t; \bar{x})$.

12. Consider the autonomous linear system $\dot{x} = ax + by$, $\dot{y} = cx + dy$, where a, b, c, d are real numbers such that $ad - bc \neq 0$. Discuss the behavior of trajectories in the neighborhood of the origin, which constitutes the unique singular point of the system.

Hint: Use the Jordan's canonical form of a 2 by 2 matrix. The following possibilities are to be considered:

$$\text{I}: \begin{pmatrix} \lambda_1 & 0 \\ 0 & \lambda_2 \end{pmatrix} \qquad \text{II}: \begin{pmatrix} \lambda & 1 \\ 0 & \lambda \end{pmatrix} \qquad \text{III}: \begin{pmatrix} \lambda & 0 \\ 0 & \bar{\lambda} \end{pmatrix}.$$

In case III, which corresponds to the existence of complex characteristic roots, the use of polar coordinates is appropriate.

13. Consider the linear control system $\dot{x} = Ax + bu(t)$, where A is a constant n by n matrix, b is a constant vector with n coordinates, and $u(t)$ is an arbitrary measurable function from $[t_0, t_1]$ into $[-1, 1]$. Assume the initial condition $x(t_0) = x^0$ is given, and denote by $K(t_1)$ the set of all points of R^n to which x^0 can be "steered" by controllers $u(t)$, as previously described. Show that $K(t_1)$ is compact and convex. Let us mention that the set $K(t_1)$ is usually called the set of *attainability*.
Hint: By definition, $\tilde{x} \in K(t_1)$ if and only if there exists a controller $u(t)$ such that $\tilde{x} = x(t_1; t_0, x^0)$. Use the formula of variation of parameters.

5

Stability Theory of Ordinary Differential Systems

From a mathematical point of view, the concept of stability can be related to that of continuous dependence of solutions on their initial values. Let us denote, as usual, by $x(t; t_0, x^0)$ the solution of the system $\dot{x} = f(t, x)$, which reduces to x^0 for $t = t_0$. Assume that $x(t; \tau, \xi)$ is defined for all $t \geq \tau$ and (τ, ξ) which belongs to a neighborhood of (t_0, x^0). If $x(t; \tau, \xi)$ is continuous at (t_0, x^0), uniformly with respect to t, $t \geq \tau$, then we shall say that $x(t; t_0, x^0)$ is a *stable* solution. The main feature of the preceding definition consists in the fact that it involves the values of the solutions on a half-axis, whereas the ordinary concept of continuous dependence is concerned with solutions defined on a finite interval (see §3.4).

On the other hand, the concept of stability has a strong physical background. To illustrate this fact, it is convenient to assume that we are dealing with the stability property of the zero solution of the system $\dot{x} = f(t, x)$ [i.e., $x(t; t_0, x^0) \equiv 0$, which implies $x^0 = 0$ and $f(t, 0) \equiv 0$]. If the system $\dot{x} = f(t, x)$ is describing the motion of a dynamical system, then our previous assumption means that this system has an equilibrium state corresponding to $x = 0$. Or, an equilibrium state is stable—from the physical point of view—if the motion of the system corresponding to small initial disturbances takes place around this state. Translated into mathematical terms, this property says that $x(t; \tau, \xi)$ is continuous at $(t_0, 0)$, uniformly with respect to t on $t \geq t_0$. In other words, the dynamical concept of stability leads directly to that of continuous dependence of solutions with respect to their initial values. The fact that we are dealing with motions taking place in an infinite period of time agrees, for instance, with such phenomena as those encountered in celestial mechanics and related fields.

There are several kinds of stability, and we shall discuss this concept in more detail. However, the reader should notice that our attempt is directed to present here only the main features of this theory, which plays an important role in contemporary research.

§5.1. Definitions and Examples

Consider the differential system

$$\dot{x} = f(t, x) \tag{5.1}$$

where x and f are vectors with n real coordinates (i.e., $x, f \in R^n$). A general assumption is the following: *$f(t, x)$ is continuous and satisfies a local Lipschitz condition in the set*

$$\Delta: \qquad t \geq 0, \qquad \|x\| < a \leq +\infty$$

and

$$f(t, 0) = 0, \qquad t \geq 0. \tag{5.2}$$

Inasmuch as our assumption implies the existence and uniqueness of the solution through any point $(t_0, x^0) \in \Delta$, it follows that $x(t; t_0, x^0)$ is defined for any $(t_0, x^0) \in \Delta$ and $t \in [t_0, T_0)$, where $T_0 \leq +\infty$ and depends on (t_0, x^0). From (5.2), it follows that $x = 0$ is a solution of (5.1), and because uniqueness holds, any solution of the system (5.1), distinct from $x = 0$, cannot vanish on its interval of existence.

We shall present here the main definitions related to the concept of stability of the solution $x = 0$ of (5.1).

Definition 1

The solution $x = 0$ of (5.1) will be called *stable* if to any $\varepsilon > 0$, $t_0 \geq 0$ there corresponds $\delta(\varepsilon, t_0) > 0$ such that $\|x(t; t_0, x^0)\| < \varepsilon$ for $t \geq t_0$, as soon as $\|x^0\| < \delta$.

Definition 2

The solution $x = 0$ of (5.1) will be called *uniformly stable* if $\delta(\varepsilon, t_0)$ from Definition 1 can be chosen independently of t_0: $\delta(\varepsilon, t_0) \equiv \delta(\varepsilon)$.

Definition 3

The solution $x = 0$ of (5.1) is called *asymptotically stable* if it is stable in the sense of Definition 1 and there exists $\gamma(t_0) > 0$ such that

$$\lim_{t \to \infty} \|x(t; t_0, x^0)\| = 0 \tag{5.3}$$

for all $x(t; t_0, x^0)$ with $\|x^0\| < \gamma$.

Definition 4

The solution $x = 0$ of (5.1) is called *uniformly asymptotically stable* if it is uniformly stable in the sense of Definition 2, and, moreover, for any $\varepsilon > 0$ there exists $T(\varepsilon) > 0$ such that

$$\|x(t; t_0, x^0)\| < \varepsilon \quad \text{for} \quad t \geq t_0 + T(\varepsilon) \tag{5.4}$$

and all x^0 with $\|x^0\| < \gamma_0$, where γ_0 is independent of t_0.

In other words, $x = 0$ is uniformly asymptotically stable if it is uniformly stable and

$$\lim \|x(t; t_0, x^0)\| = 0 \quad \text{as} \quad t - t_0 \to +\infty$$

uniformly with respect to (t_0, x^0), $t_0 \geq 0$, $\|x^0\| < \gamma_0$. Concerning the relationships among these concepts of stability, it is obvious that uniform stability and asymptotic stability imply stability in the sense of Definition 1 and that uniform asymptotic stability implies all other kinds of stability.

Before proceeding further with our investigation, we shall discuss the stability properties of the scalar equation

$$\dot{x} = a(t)x \tag{5.5}$$

where $a(t)$ is a continuous function on the half-axis $t \geq 0$. From (5.5), one obtains

$$x(t; t_0, x^0) = x^0 \exp\left[\int_{t_0}^{t} a(s)\, ds\right]. \tag{5.6}$$

The following statements hold:

1. The solution $x = 0$ of (5.5) is stable if and only if

$$\int_{t_0}^{t} a(s)\, ds \leq M(t_0), \qquad t \geq t_0 \tag{5.7}$$

where $M(t_0)$ is finite for any $t_0 \geq 0$.

2. The solution $x = 0$ of (5.5) is uniformly stable if and only if

$$\int_{t_0}^{t} a(s)\, ds \leq M < \infty, \qquad t \geq t_0 \geq 0 \tag{5.8}$$

with M constant.

3. The solution $x = 0$ of (5.5) is asymptotically stable if and only if

$$\lim_{t \to \infty} \int_{t_0}^{t} a(s)\, ds = -\infty. \tag{5.9}$$

4. The solution $x = 0$ of (5.5) is uniformly asymptotically stable if and only if there exists $\alpha > 0$, such that

$$\int_{t_0}^{t} a(s)\, ds \leq -\alpha(t - t_0), \qquad t \geq t_0 \geq 0. \tag{5.10}$$

The proofs of the first three statements are relatively simple. Concerning statement 4, we shall give the proof in the next section, when we discuss stability properties of linear systems (actually, we shall be concerned with a more general case).

Let us now prove statement 1. If (5.7) holds, then (5.6) gives

$$|x(t; t_0, x^0)| \leq |x^0| \exp\{M(t_0)\} < \varepsilon, \quad \text{for} \quad t \geq t_0$$

as soon as $|x^0| < \delta\ (\varepsilon, t_0) = \varepsilon \exp\{-M(t_0)\}$. Hence, (5.7) is sufficient for stability. To prove its necessity, assume it is violated for some $t_0 > 0$. Then we can find a sequence $\{t_n\}$, $n \geq 1$, such that $t_n \to \infty$ and

$$\int_{t_0}^{t_n} a(s)\, ds > \log n, \qquad n = 1, 2, \ldots. \tag{5.11}$$

This means that $|x(t_n; t_0, x^0)| > n\,|x^0|$, which shows that $x = 0$ cannot be stable for (5.5).

The proof of statement 2 is quite similar. In particular, one can take $\delta(\varepsilon) = \varepsilon \exp\{-M\}$.

Consider now assertion 3. Inasmuch as (5.9) implies (5.7), it follows that (5.9) implies the stability. Moreover, (5.9) implies

$$\lim_{t \to \infty} x(t; t_0, x^0) = 0 \tag{5.12}$$

no matter how we take $t_0 \geq 0$ and x^0 real. Consequently, (5.9) implies the asymptotic stability.

To prove the necessity of (5.9), assume that $x = 0$ is asymptotically stable. If (5.9) were not true, then we could find a sequence $\{t_n\}$, $t_n \to \infty$ as $n \to \infty$, and a number $N > 0$ such that

$$\int_{t_0}^{t_n} a(s)\, ds \geq \log N > -\infty, \qquad n = 1, 2, \ldots. \tag{5.13}$$

The preceding inequality implies $|x(t_n; t_0, x^0)| \geq N|x^0|$, which contradicts the asymptotic stability property of the solution $x = 0$ of (5.5).

From the preceding statements we can conclude that the four kinds of stability are all distinct, because no pair of conditions (5.7), (5.8), (5.9), (5.10) consists of equivalent conditions. The following scheme is useful:

$$\begin{array}{ccccc} & & \text{uniform stability} & & \\ & \nearrow & & \searrow & \\ \text{uniform asymptotic stability} & & & & \text{stability.} \\ & \searrow & & \nearrow & \\ & & \text{asymptotic stability} & & \end{array}$$

Concerning the relationship between uniform stability and asymptotic stability, we shall remark that condition (5.8) is satisfied by $a(t) = \cos t$, whereas (5.9) is not. Analogously, $a(t) = \sin \log t + \cos \log t - \alpha$, $1 < \alpha < \sqrt{2}$, satisfies (5.10) (for $t_0 = 1$) but does not satisfy (5.8). Indeed, $\int_1^t a(s)\,ds = t \sin \log t - \alpha(t - 1)$, from which (5.10) follows immediately. On the other hand, $\sin x + \cos x \geq \beta > \alpha$ for $x \in [x_1, x_2]$, where $0 < x_1 < \pi/4 < x_2 < \pi/2$. Hence, for $\tau_n = \exp(2n\pi + x_1) < t_n = \exp(2n\pi + x_2)$, we obtain

$$\int_{\tau_n}^{t_n} a(s)\,ds \geq (\beta - \alpha)e^{2n\pi}[\exp(x_2) - \exp(x_1)] \to \infty \quad \text{as} \quad n \to \infty.$$

Consequently, uniform stability and asymptotic stability are distinct concepts.

A few comments are necessary before we proceed further with our discussion. First, in defining the various concepts of stability we discussed such kinds of properties only for the solution $x = 0$ of the system (5.1). The corresponding properties for an arbitrary solution $x = x(t)$ of the system (5.1) can be reduced to the preceding case.

Indeed, if $x = x(t)$ is a solution of (5.1) defined on the half-axis $t \geq 0$, then $y = x - x(t)$ satisfies

$$\dot{y} = g(t, y) \tag{5.14}$$

where $g(t, y) = f(t, y + x(t)) - f(t, x(t))$. We have $g(t, 0) \equiv 0$, and from Definition 1 we derive the following: *The solution $x = x(t)$ of (5.1) is stable if to any $\varepsilon > 0$, $t_0 \geq 0$, there corresponds $\delta(\varepsilon, t_0) > 0$ such that*

$$\|x(t; t_0, x^0) - x(t)\| < \varepsilon,$$

whenever $t \geq t_0$ and $\|x^0 - x(t_0)\| < \delta$.

Similar considerations can be made with respect to other types of stability. From a dynamical point of view, we can say that the stability problem of a given motion can be reduced to the stability of an equilibrium state (for an appropriate dynamical system).

Another useful remark is concerned with the role of t_0 in Definitions 1 through 4. It is obvious that this variable plays an essential role in Definitions 2 and 4. On the contrary, in Definitions 1 and 3 we can refer to a fixed (unique) value of t_0—say, $t_0 = 0$—to obtain the same concept of stability.

Indeed, assume the property stated in Definition 1 holds for $t_0 = 0$. Let us prove that it holds also for an arbitrary $t_0 \geq 0$, using the continuous dependence of solutions on their initial values (on any finite interval!). Consider an arbitrary $\varepsilon > 0$ and let $t_0 > 0$ be given. There exists $\eta(\varepsilon) > 0$ such that $t \geq 0$ and $\|x^0\| < \eta$ imply $\|x(t; 0, x^0)\| < \varepsilon$. But $x^0 = x(0; t_0, \tilde{x}^0)$, and $\|x^0\| < \eta$ is implied by $\|\tilde{x}^0\| < \delta(\varepsilon, t_0)$. More precisely, $\|x(t; t_0, \tilde{x}^0)\| < \eta(\varepsilon)$ for $t \in [0, t_0]$, as soon as $\|\tilde{x}^0\| < \delta(\varepsilon, t_0)$. Consequently, $\|x(t; t_0, \tilde{x}^0)\| < \varepsilon$ for $t \geq 0$ (hence, for $t \geq t_0$) and $\|\tilde{x}^0\| < \delta(\varepsilon, t_0)$, because $x(t; t_0, \tilde{x}^0) = x(t; 0, x^0)$ [see (3.19) in §3.4].

Finally, we would like to point out that several other concepts of stability can be defined. For instance, the concept of *exponential asymptotic stability* will appear naturally in the case of linear systems. Also, we shall deal with systems that are *globally asymptotically stable* (or stable in the large). These concepts will be considered in the sequel.

§5.2. Stability of Linear Systems

Consider the homogeneous linear system

$$\dot{x} = A(t)x \tag{5.15}$$

where $A(t)$ is a continuous n by n matrix on the half-axis $t \geq 0$. We know that

$$x(t; t_0, x^0) = X(t)X^{-1}(t_0)x^0, \qquad t \geq 0 \tag{5.16}$$

where $X(t)$ is an arbitrary fundamental matrix of the system (5.15) (see §4.2) defined on the whole half-axis $t \geq 0$.

Let us consider the following problem: Find necessary and sufficient conditions for $X(t)$ that ensure a given kind of stability for the solution $x = 0$ of the system (5.15). We consider first the cases of stability and asymptotic stability.

Theorem 5.1

Let $X(t)$ be a fundamental matrix of the system (5.15). *A necessary and sufficient condition for the stability of the solution $x = 0$ is the boundedness of $X(t)$ on $t \geq 0$:*

$$\|X(t)\| \leq M, \qquad t \geq 0. \tag{5.17}$$

A necessary and sufficient condition for the asymptotic stability of the solution $x = 0$ *is*

$$\lim_{t\to\infty} \|X(t)\| = 0. \tag{5.18}$$

Proof. If (5.17) holds, then (5.16) leads to

$$\|x(t;\, t_0,\, x^0)\| \leq M \,\|X^{-1}(t_0)\| \cdot \|x^0\|, \qquad t \geq 0 \tag{5.19}$$

from which we obtain $\|x(t;\, t_0,\, x^0)\| < \varepsilon$, as soon as $\|x^0\| < \delta(\varepsilon, t_0) = \varepsilon(M\|X^{-1}(t_0)\|)^{-1}$. Hence, condition (5.17) is sufficient for stability.

To prove the necessity of (5.17), it suffices to take into account the fact that all the solutions of (5.15) are bounded if the zero solution is stable. Indeed, we shall obtain $\|x(t;\, 0,\, x^0)\| < 1$ on $t \geq 0$, for $\|x^0\| < \delta$. In other words, all the solutions of (5.15), with $\|x^0\| < \delta$, are bounded. From the linearity it follows easily that any solution of (5.15) is bounded on $t \geq 0$. Inasmuch as any column of $X(t)$ is a solution of (5.15), it follows that $X(t)$ is bounded on $t \geq 0$. The preceding arguments show also that stability and boundedness are equivalent properties for the system (5.15).

Consider now the case of asymptotic stability of the solution $x = 0$ of (5.15). Inasmuch as (5.18) implies (5.17), (5.18) implies the stability. Moreover, from (5.16) we get

$$\|x(t;\, t_0,\, x^0)\| \leq \|X(t)\| \cdot \|X^{-1}(t_0)\| \;\|x^0\|, \qquad t \geq 0 \tag{5.20}$$

which implies $\lim\|x(t;\, t_0,\, x^0)\| = 0$ as $t \to \infty$ for any solution of the system (5.15). Hence, (5.18) is a sufficient condition for the asymptotic stability.

To prove the necessity of (5.18), we shall remark that

$$\lim_{t\to\infty} \|x(t;\, 0,\, x^0)\| = 0 \tag{5.21}$$

holds for all x^0 such that $\|x^0\| < \delta_0$. This property implies (5.21) for all x^0, as follows easily from the linearity. In particular, all the solutions forming the columns of $X(t)$ tend toward zero as $t \to \infty$—that is, (5.18) holds. Theorem 5.1 is thus proved.

The next result is concerned with uniform and uniform asymptotic stability.

Theorem 5.2

Consider the system (5.15) *and let* $X(t)$ *be a fundamental matrix of this system. A necessary and sufficient condition for the uniform stability of the solution* $x = 0$ *is the existence of a number* $M > 0$ *such that*

$$\|X(t)X^{-1}(t_0)\| \leq M, \qquad t \geq t_0 \geq 0. \tag{5.22}$$

A necessary and sufficient condition for the uniform asymptotic stability of the solution $x = 0$ is the existence of two positive numbers N and α, such that

$$\|X(t)X^{-1}(t_0)\| \leq N \exp[-\alpha(t - t_0)], \qquad t \geq t_0 \geq 0. \tag{5.23}$$

Proof. The sufficiency of conditions (5.22) and (5.23) can be easily verified if we consider (5.16). We obtain $\|x(t; t_0, x^0)\| \leq M\|x^0\|$, which shows that $\|x(t; t_0, x^0)\| < \varepsilon$ for $t \geq t_0$, if $\|x^0\| < \varepsilon M^{-1}$. Similarly, from (5.16) and (5.23) we get $\|x(t; t_0, x^0)\| < \varepsilon$ for $\|x^0\| < 1$ and

$$t \geq t_0 + \alpha^{-1}\log(\varepsilon^{-1}N) = t_0 + T(\varepsilon).$$

Let us prove now the necessity of (5.22). If $x = 0$ is uniformly stable, then

$$\|X(t)X^{-1}(t_0)x^0\| < 1 \quad \text{for} \quad \|x^0\| \leq \delta \tag{5.24}$$

no matter how we take t and t_0, with $t \geq t_0 \geq 0$. Condition (5.24) is equivalent to

$$\|X(t)X^{-1}(t_0)x^0\| < \delta^{-1} \quad \text{for} \quad \|x^0\| \leq 1. \tag{5.25}$$

From (5.25) we can find easily

$$\|X(t)X^{-1}(t_0)\| < \sqrt{n}\,\delta^{-1}, \qquad t \geq t_0 \geq 0 \tag{5.26}$$

where n is the number of rows and columns for $X(t)$. Indeed, if we choose in (5.25) x^0 such that $\|x^0\| = 1$ and all the coordinates are zero, except that of rank m, $1 \leq m \leq n$, then we obtain $\|\mathrm{col}_m\, X(t)X^{-1}(t_0)\| < \delta^{-1}$. These inequalities lead immediately to (5.26), which shows that (5.22) holds with $M = \sqrt{n}\,\delta^{-1}$.

Finally, we have to prove the necessity of condition (5.23). If $x = 0$ is uniformly asymptotically stable, there then exists a $\gamma_0 > 0$, and for each $\varepsilon > 0$ there corresponds $T(\varepsilon)$, such that $\|x^0\| < \gamma_0$ and $t \geq t_0 + T(\varepsilon)$ imply

$$\|X(t)X^{-1}(t_0)x^0\| < \varepsilon. \tag{5.27}$$

This inequality yields [see the argument developed in proving the necessity of (5.22)]

$$\|X(t)X^{-1}(t_0)\| < k \quad \text{for} \quad t \geq t_0 + T \tag{5.28}$$

where $k = \sqrt{n}\,\gamma_0^{-1}\varepsilon < 1$. Inequality (5.28) can be written also in the form

$$\|X(t_0 + T)X^{-1}(t_0)\| < k, \qquad t_0 \geq 0. \tag{5.29}$$

On the other hand, inasmuch as uniform asymptotic stability implies uniform stability, there exists $A > 0$ such that

$$\|X(t_0 + s)X^{-1}(t_0)\| \leq A, \qquad t_0 \geq 0, \qquad 0 \leq s \leq T. \tag{5.30}$$

From (5.29) and (5.30), we shall derive (5.23). Indeed, for any $t \geq t_0$ there exists a natural m, such that $t \in [t_0 + mT, t_0 + (m+1)T)$. Consequently,

$$\|X(t)X^{-1}(t_0)\| \leq \|X(t)X^{-1}(t_0 + mT)\| \cdot \|X(t_0 + mT)X^{-1}(t_0 + \overline{m-1\;T})\| \cdots \|X(t_0 + T)X^{-1}(t_0)\| < Ak^m.$$

Denote $\alpha = -T^{-1} \log k$, which implies $\alpha > 0$ and

$$\|X(t)X^{-1}(t_0)\| < k^{-1}A \exp[-\alpha(m+1)T] \leq N \exp[-\alpha(t - t_0)] \quad (5.31)$$

for $t \geq t_0$, if we set $k^{-1}A = N$ and take into account that $t - t_0 < (m+1)T$. From (5.31) we obtain (5.23), considering the extreme terms. Theorem 5.2 is completely proved.

Remark. It follows from the preceding theorem that the uniform asymptotic stability of the zero solution of (5.15) implies the exponential decay at infinity of all the solutions of (5.15). Namely, from (5.16) and (5.23) we get

$$\|x(t; t_0, x^0)\| \leq N \|x^0\| \exp[-\alpha(t - t_0)], \qquad t \geq t_0 \quad (5.32)$$

which proves the statement.

If the system $\dot{x} = f(t, x)$ is such that (5.32) holds for all its solutions with $\|x^0\|$ sufficiently small (say, $\|x^0\| \leq \gamma$), then $x = 0$ is called *exponentially asymptotically stable*. Of course, the exponential asymptotic stability implies always the uniform asymptotic stability. The converse is also true for linear systems, as we have seen in Theorem 5.2.

In concluding this section, we shall apply the preceding results to the case of systems with constant coefficients:

$$\dot{x} = Ax. \quad (5.33)$$

These systems are of the form (5.15), with $A(t) \equiv A$.

We first want to state that the concept of uniform stability coincides with that of stability, and, similarly, uniform asymptotic stability is the same as asymptotic stability. Indeed, if $X(t)$ is the fundamental matrix of (5.33), such that $X(0) = I$, then

$$X(t)X^{-1}(t_0) = X(t - t_0). \quad (5.34)$$

This follows easily from the fact that both sides (5.34) are solutions of $\dot{X} = AX$, which reduce to I for $t = t_0$. It can be also derived from $X(t) = \exp(At)$. Taking into account (5.34), (5.16) can be written as

$$x(t; t_0, x^0) = X(t - t_0)x^0 \quad (5.35)$$

which leads easily to the equivalence previously mentioned.

For instance, the asymptotic stability of $x = 0$ implies (5.18)—that is, to any $\varepsilon > 0$ there corresponds $T(\varepsilon) > 0$ such that $\|X(t)\| < \varepsilon$ for $t \geq T(\varepsilon)$. From (5.35) we obtain

$$\|x(t; t_0, x^0)\| < \varepsilon \quad \text{for} \quad t \geq t_0 + T(\varepsilon)$$

if $\|x^0\| \leq 1 \ (= \gamma_0)$. Hence, the asymptotic stability implies the uniform asymptotic stability—that is, the exponential asymptotic stability.

Starting from the preceding results and considering what we found in §4.6, we can derive the following conclusions.

The zero solution of (5.33) *is stable* (*hence, uniformly stable*), *if and only if all the characteristic roots of the matrix A have nonpositive real parts and those with zero real parts are such that the corresponding blocks in the Jordan canonical form reduce to a single element* (1 *by* 1 *matrix*).

Another way to express the last property in the preceding statement is to say that all the characteristic roots of A, with zero real parts, have simple elementary divisors. Indeed, only in this case are the solutions of (5.33) all bounded on $t \geq 0$. The only solutions of the form $\xi t^k e^{i\alpha t}$, where α is a real number and ξ is a vector, correspond to $k = 0$.

The zero solution of (5.33) *is asymptotically stable* (*hence, exponentially asymptotically stable*), *if and only if all the characteristic roots of A have negative real parts.*

Indeed, $X(t)$ will tend toward zero as $t \to \infty$ if and only if all the quasi-polynomials constituting it have their exponents with negative real parts. Inasmuch as the exponents of these quasi-polynomials are nothing else but the characteristic roots of A, the statement is proved.

Consequently, both criteria are concerned with the problem of finding the signs of the real parts of the characteristic roots of the matrix A. There exist several conditions that ensure the fact that all the roots of an algebraic equation

$$a_0 \lambda^n + a_1 \lambda^{n-1} + \cdots + a_{n-1}\lambda + a_n = 0, \qquad a_0 > 0 \tag{5.36}$$

have negative real parts. We shall mention here the well-known criterion of Hurwitz.[1]

Let us form the n by n matrix

$$\begin{pmatrix} a_1 & a_3 & a_5 & \cdots & 0 \\ a_0 & a_2 & a_4 & \cdots & 0 \\ 0 & a_1 & a_3 & \cdots & 0 \\ 0 & a_0 & a_2 & \cdots & 0 \\ & & & \vdots & \\ 0 & & & \cdots & a_{n-2}\, a_n \end{pmatrix}$$

and denote by Δ_k its principal *minor* of order k, $1 \leq k \leq n$.

[1] See J. V. Uspensky, *Theory of Equations*, McGraw-Hill Book Company, New York, 1948, p. 304.

A necessary and sufficient condition for Eq. (5.36) *to have all its roots with negative real parts is* $\Delta_k > 0$, $k = 1, 2, \ldots, n$.

For instance, in the case of $n = 3$, we find

$$\Delta_1 = a_1 > 0, \qquad \Delta_2 = \begin{vmatrix} a_1 & a_3 \\ a_0 & a_2 \end{vmatrix} > 0, \qquad \Delta_3 = a_3 \Delta_2 > 0.$$

§5.3. Stability in the First Approximation

Beside our linear system (5.15), let us consider the nonlinear system

$$\dot{x} = A(t)x + f(t, x), \qquad f(t, 0) \equiv 0 \tag{5.37}$$

where $f(t, x)$ is a continuous vector function in the set,

$$\Delta: \qquad t \geq 0, \qquad \|x\| < a \leq +\infty$$

satisfying a local Lipschitz condition.

System (5.37) is usually called the *perturbed system* associated with (5.15), whereas (5.15) is called the *system of the first approximation* corresponding to (5.37). This terminology is related to the fact that $f(t, x)$ is "small," in a sense that should be made more precise. Inasmuch as our assumption $f(t, 0) \equiv 0$ shows that (5.37) possesses the zero solution, it is natural to investigate the stability properties of this solution, starting from the fact that the system of the first approximation has such properties. The first result we shall establish is concerned with the uniform stability.

Theorem 5.3

Consider the system (5.37) *where* $A(t)$ *is such that the zero solution of the system in the first approximation* (5.15) *is uniformly stable. Assume that*

$$\|f(t, x)\| \leq g(t)\|x\| \tag{5.38}$$

in Δ, *where* $g(t)$ *is a continuous function on* $t \geq 0$, *such that*

$$\int_0^\infty g(s)\, ds < +\infty. \tag{5.39}$$

Then the solution $x = 0$ *of* (5.37) *is uniformly stable.*

Proof. First, we shall prove that system (5.37) with the initial condition $x(t_0) = x^0$ is equivalent to

$$x(t; t_0, x^0) = X(t)X^{-1}(t_0)x^0 + \int_{t_0}^t X(t)X^{-1}(s)f(s, x(s; t_0, x^0))\, ds \tag{5.40}$$

which constitutes a vector integral equation.

Indeed, from (5.37) we obtain (5.40) by applying the formula of variation of constants (see §4.3). Conversely, from (5.40) we get, by differentiation with respect to t,

$$\frac{d}{dt}x(t; t_0, x^0) = \frac{dX(t)}{dt}X^{-1}(t_0)x^0 + f(t; x(t; t_0, x^0)) + \int_{t_0}^{t} \frac{dX(t)}{dt} X^{-1}(s)f(s, x(s; t_0, x^0))\, ds$$

which leads immediately to

$$\frac{d}{dt}x(t; t_0, x^0) = A(t)x(t; t_0, x^0) + f(t, x(t; t_0, x^0))$$

if we consider that $X(t)$ is a matrix solution for $\dot{X} = A(t)X$. Consequently, we can deal with (5.40) instead of (5.37).

Inasmuch as the solution $x = 0$ of (5.15) is uniformly stable, the fundamental matrix $X(t)$ will satisfy the estimate (5.22). Hence,

$$\|x(t; t_0, x^0)\| \leq M\,\|x^0\| + M \int_{t_0}^{t} \|f(s, x(s; t_0, x^0))\|\, ds, \qquad t \geq t_0.$$

But $f(t, x)$ satisfies (5.38), which yields

$$\|x(t; t_0, x^0)\| \leq M\,\|x^0\| + M \int_{t_0}^{t} g(s)\,\|x(s; t_0, x^0)\|\, ds, \qquad t \geq t_0. \tag{5.41}$$

Inequality (5.41) is valid for those values of $t \geq t_0$ for which $x(t; t_0, x^0)$ is defined—that is, $\|x(t; t_0, x^0)\| < a$. We shall see that this is true for all $t \geq t_0$, as soon as $\|x^0\|$ is sufficiently small.

Inasmuch as (5.41) is an inequality of Gronwall type, we obtain

$$\|x(t; t_0, x^0)\| \leq M\,\|x^0\| \exp\left\{M \int_{t_0}^{t} g(s)\, ds\right\}, \qquad t \geq t_0. \tag{5.42}$$

Taking into account condition (5.39), it follows that $\|x(t; t_0, x^0)\| < a$ for $t \geq t_0$, as soon as $M\|x^0\| \exp\{M \int_0^\infty g(s)\, ds\} < a$. The uniform stability of the solution $x = 0$ of (5.37) is now the consequence of the inequality

$$\|x(t; t_0, x^0)\| \leq M\,\|x^0\| \exp\left\{M \int_{0}^{\infty} g(s)\, ds\right\}, \qquad t \geq t_0 \tag{5.43}$$

which is an obvious implication of (5.42).

Corollary

Consider the system

$$\dot{x} = [A + B(t)]x \tag{5.44}$$

where x is an n-vector and A, B are n by n matrices. Assume A is a constant matrix such that all the solutions of the system $\dot{x} = Ax$ are bounded on $t \geq 0$ (hence, $x = 0$ is uniformly stable). If $B(t)$ is such that

$$\int_0^\infty \|B(t)\| \, dt < +\infty \tag{5.45}$$

then $x = 0$ is uniformly stable for (5.44). In particular, it follows that if we perturb the system $\dot{x} = Ax$, the solutions of which are bounded on $t \geq 0$, by adding the term $B(t)x$, with $B(t)$ satisfying (5.45), the boundedness property remains valid.

Let us consider now the case when the zero solution of the system in the first approximation is uniformly asymptotically stable (hence, exponentially asymptotically stable). The result we shall prove is known as Poincaré–Liapunov theorem.

Theorem 5.4

Consider system (5.37) *under the following assumptions*:

1. $A(t)$ *is such that the solution* $x = 0$ *of* (5.15) *is exponentially asymptotically stable*;
2. $f(t, x)$ *satisfies the estimate*

$$\|f(t, x)\| \leq L \|x\| \tag{5.46}$$

where $L > 0$ *is sufficiently small. Then* $x = 0$ *is exponentially asymptotically stable for* (5.37).

Proof. We shall imitate the proof of Theorem 5.3. Inasmuch as (5.37) with the initial condition $x(t_0) = x^0$ is equivalent to (5.40), we find from (5.40), (5.23), and (5.46),

$$\|x(t; t_0, x^0)\| \leq N \|x^0\| \exp[-\alpha(t - t_0)] + LN \int_{t_0}^{t} e^{-\alpha(t-s)} \|x(s; t_0, x^0)\| \, ds \qquad t \geq t_0. \tag{5.47}$$

If we multiply both sides of (5.47) by $e^{\alpha t}$ and denote $e^{\alpha t}\|x(t; t_0, x^0)\| = u(t)$, we get

$$u(t) \leq N \|x^0\| \exp(\alpha t_0) + LN \int_{t_0}^{t} u(s) \, ds, \qquad t \geq t_0. \tag{5.48}$$

Inequality (5.48) is of Gronwall type, and we obtain

$$e^{\alpha t}\|x(t; t_0, x^0)\| \leq N \|x^0\| \exp(\alpha t_0) \exp[LN(t - t_0)], \qquad t \geq t_0. \tag{5.49}$$

Assume now that $LN < \alpha$—that is,

$$L < \alpha N^{-1}. \tag{5.50}$$

Inequality (5.50) shows how small L should be chosen to ensure the exponential asymptotic stability of the solution $x = 0$ of (5.37). Inequality (5.49) can be written

$$\|x(t; t_0, x^0)\| \leq N \|x^0\| \exp[-(\alpha - LN)(t - t_0)], \qquad t \geq t_0. \tag{5.51}$$

It holds for all $t \geq t_0$ such that $x(t; t_0, x^0)$ has its graph in Δ. Consequently, Eq. (5.51) is valid for all $t \geq t_0$, as soon as $\|x^0\|$ is sufficiently small ($N \|x^0\| < a$). From (5.51), it follows that $x = 0$ is an exponentially asymptotically stable solution of the system (5.37). Theorem 5.4 is completely proved.

Corollary

Consider the system

$$\dot{x} = Ax + f(t, x) \tag{5.52}$$

where A has all its characteristic roots with negative real parts and $f(t, x)$ satisfies the general assumption of this section and condition 2 from Theorem 5.4. Then the solution $x = 0$ of (5.52) is exponentially asymptotically stable.

A noteworthy particular case of the corollary is obtained for $f(t, x) \equiv f(x)$, where $f(x)$ is such that its coordinates are power series in $x_1, x_2, \ldots, x_n$, starting with quadratic terms. If A is as described, then $x = 0$ is exponentially asymptotically stable for the system $\dot{x} = Ax + f(x)$. Indeed, the derivatives $\partial f_i/\partial x_j$ tend toward zero as $\|x\| \to 0$, which means that a Lipschitz condition holds. The Lipschitz constant is arbitrarily small, if we choose the number a small enough in defining Δ.

Consequently, if the second members of a differential system are analytic functions in a neighborhood of $x = 0$ (which is assumed to be an equilibrium point), then the system in the first approximation offers simple conditions for asymptotic stability.

§5.4. Stability Theorems by Comparison Method

Let us consider again the system (5.1)

$$\dot{x} = f(t, x), \qquad f(t, 0) \equiv 0$$

under the same general assumption concerning $f(t, x)$: continuity in Δ and the local Lipschitz condition with respect to x. To obtain efficient criteria of stability for the solution $x = 0$ of (5.1), we shall use now the comparison method. The core of this method, as told in §3.5, consists in the simultaneous use of the so-called Liapunov functions and the differential inequalities.

We shall consider now some properties related to the concept of Liapunov function. Assume that $\lambda(r)$, $\mu(r)$, ... are increasing continuous functions of the real variable r, $r \in [0, a)$ a being the same as in the definition of Δ, such that they vanish at $r = 0$. Consequently, they are positive for $r > 0$.

Let us denote by $K = K[0, a)$ the class of functions with the properties indicated. If $V(t, x)$ is a real-valued function defined on

$$\Delta = \{(t, x); t \geq 0, \|x\| < a \leq +\infty\}$$

we shall say that $V(t, x)$ is *positive definite* if $V(t, 0) = 0$ for $t \geq 0$ and if there exists a function $\lambda(r) \in K$, such that $\lambda(\|x\|) \leq V(t, x)$ on Δ. A function $V(t, x)$ will be called *decrescent* if there exists a function $\mu(r) \in K$ such that $V(t, x) \leq \mu(\|x\|)$, $(t, x) \in \Delta$. The meaning of the term *negative definite* is obvious.

Assume now that $V(t, x)$ is differentiable on Δ and satisfies an inequality of the form

$$\frac{\partial V}{\partial t} + \left(\frac{\partial V}{\partial x}, f\right) \leq \omega(t, V) \tag{5.53}$$

as required by the comparison principle. As mentioned in §3.5,

$$\left(\frac{\partial V}{\partial x}, f\right) = \frac{\partial V}{\partial x_1} f_1 + \frac{\partial V}{\partial x_2} f_2 + \cdots + \frac{\partial V}{\partial x_n} f_n.$$

Concerning $\omega(t, y)$, which determines the comparison equation

$$\dot{y} = \omega(t, y) \tag{5.54}$$

we shall assume that it is continuous on the set $t \geq 0, 0 \leq y < A \leq +\infty$ and satisfies a local Lipschitz condition with respect to y. Moreover, $\omega(t, 0) \equiv 0$.

We are able to prove the main result concerning the stability by the comparison method.

Theorem 5.5

Consider the system (5.1) *under the general assumption from* §5.1 *and assume there exists a function* $V(t, x)$ *that satisfies* (5.53) *on* Δ.

If $V(t, x)$ *is positive definite and the solution* $y = 0$ *of* (5.54) *is stable* (*asymptotically stable*), *then the solution* $x = 0$ *of* (5.1) *is also stable* (*asymptotically stable*).

If $V(t, x)$ is positive definite and decrescent and the solution $y = 0$ of (5.54) is uniformly stable (uniformly asymptotically stable), then the solution $x = 0$ of (5.1) is also uniformly stable (uniformly asymptotically stable).

Proof. If $(t_0, x^0) \in \Delta$, then $x(t; t_0, x^0)$ is defined on $[t_0, T)$, $T \leq +\infty$ and we have $(t, x(t; t_0, x^0)) \in \Delta$ for $t \in [t_0, T)$.

As shown in §3.5, inequality (5.53) implies

$$V(t, x(t; t_0, x^0)) \leq y(t; t_0, y_0), \qquad t \in [t_0, T) \tag{5.55}$$

where $y(t; t_0, y_0)$ is the solution of the comparison equation (5.54) with

$$V(t_0, x^0) \leq y_0. \tag{5.56}$$

Now let $\varepsilon > 0$ be an arbitrary number ($\varepsilon < a$). If $\lambda(\|x\|) \leq V(t, x)$, $(t, x) \in \Delta$, then our hypothesis allows us to write

$$y(t; t_0, y_0) < \lambda(\varepsilon), \qquad t \geq t_0 \tag{5.57}$$

as soon as

$$y_0 < \eta(\varepsilon, t_0). \tag{5.58}$$

Consequently, one obtains

$$\lambda(\|x(t; t_0, x^0)\|) < \lambda(\varepsilon), \qquad t \in [t_0, T)$$

which yields

$$\|x(t; t_0, x^0)\| < \varepsilon, \qquad t \in [t_0, T) \tag{5.59}$$

for all x^0 satisfying $V(t_0, x^0) < \frac{1}{2}\eta(\varepsilon, t_0)$. In other words, (5.59) holds for all x^0 such that

$$\|x^0\| < \delta(\varepsilon, t_0). \tag{5.60}$$

We shall prove now that (5.59) implies $T = +\infty$—that is, a solution $x(t; t_0, x^0)$ that satisfies (5.59) on its maximal interval of existence is necessarily defined on the half-axis $t \geq t_0$. Indeed, from $\varepsilon < a$ it follows that a solution $x(t; t_0, x^0)$ with $T < +\infty$, will leave the half-cylinder $t \geq t_0$, $\|x\| < \varepsilon$, for some $t_1 > t_0$, with $t_0 < t_1 < T$. Hence, $\|x(t_1, t_0, x^0)\| > \varepsilon$, which contradicts (5.59).

The preceding considerations show that for any $\varepsilon > 0$, $t_0 \geq 0$, there exists $\delta(\varepsilon, t_0)$ such that (5.60) implies (5.59), with $T = +\infty$. This proves that the stability of the solution $y = 0$ of the comparison equation (5.54) and the existence of a positive-definite function $V(t, x)$ satisfying (5.53) imply the stability of the solution $x = 0$ of the system (5.1). Concerning the asymptotic stability, it is sufficient to remark that

$$\lambda(\|x(t; t_0, x^0)\|) \leq y(t; t_0, V(t_0, x^0)) \tag{5.61}$$

holds for $t \geq t_0$ and $\|x^0\|$ small enough, which implies

$$\lim_{t \to \infty} \|x(t; t_0, x^0)\| = 0.$$

Consider now the case of the uniform stability. The only point of the proof needing further consideration is concerned with the choice of $\delta(\varepsilon, t_0)$. In the case of uniform stability, we know that $V(t, x) \leq \mu(\|x\|)$ and $\eta(\varepsilon, t_0)$ can be chosen independently of t_0: $\eta(\varepsilon, t_0) \equiv \eta(\varepsilon)$. If we choose $\delta(\varepsilon)$ such that $\mu(\delta) < \frac{1}{2}\eta(\varepsilon)$—that is, $\delta < \mu^{-1}(\frac{1}{2}\eta(\varepsilon))$, then we obtain

$$\|x(t; t_0, x^0)\| < \varepsilon \quad \text{for} \quad t \geq t_0 \quad \text{and} \quad \|x^0\| < \delta(\varepsilon). \tag{5.62}$$

This proves that the uniform stability of the solution $y = 0$ of (5.54) and the existence of a function $V(t, x)$, which is positive definite and decrescent, imply the uniform stability of the solution $x = 0$ of the system (5.1).

Finally, to prove the assertion related to the uniform asymptotic stability, we shall first point out that the hypotheses imply uniform stability. Moreover, for any $\varepsilon > 0$, there exists $T(\varepsilon) > 0$ such that

$$y(t; t_0, y_0) < \lambda(\varepsilon), \qquad t \geq t_0 + T(\varepsilon) \tag{5.63}$$

where $y_0 < \eta_0$, η_0 being a fixed number. Inasmuch as $V(t, x) \leq \mu(\|x\|)$, if we choose δ_0 such that $\mu(\delta_0) < \eta_0$, then (5.63) implies

$$\|x(t; t_0, x^0)\| < \varepsilon, \qquad t \geq t_0 + T(\varepsilon) \tag{5.64}$$

for all x^0 such that $\|x^0\| < \delta_0$. This proves the uniform asymptotic stability of the solution $x = 0$ of (5.1) and Theorem 5.5 is thus proved. Several classical results on stability can be derived from the preceding theorem.

Corollary

(*First Liapunov's theorem on stability.*) *If there exists a positive definite function $V(t, x)$ on Δ, such that*

$$\frac{\partial V}{\partial t} + \left(\frac{\partial V}{\partial x}, f\right) \leq 0 \tag{5.65}$$

then $x = 0$ is a stable solution of (5.1).

Indeed, for $\omega(t, y) \equiv 0$ we obtain $y(t; t_0, y_0) = y_0$, which shows that $y = 0$ is stable.

Corollary

(*Persidski's theorem on uniform stability.*) *Assume that there exists a positive-definite decrescent function $V(t, x)$, such that* (5.65) *holds. Then $x = 0$ is uniformly stable.*

This statement follows from Theorem 5.5, if we take into account the fact that $y = 0$ is uniformly stable for the comparison equation $\dot{y} = 0$.

Corollary

(*Liapunov's second theorem on stability.*) This result is actually concerned with uniform asymptotic stability. *Assume that there exists a positive definite decrescent function* $V(t, x)$ *such that its derivative with respect to* (5.1) *is negative definite. Then the solution* $x = 0$ *of* (5.1) *is uniformly asymptotically stable.*

The proof of the preceding statement requires further arguments. Let us remark that $V(t, x)$ satisfies the following conditions:

$$\lambda(\|x\|) \leq V(t, x) \leq \mu(\|x\|) \tag{5.66}$$

$$\frac{\partial V}{\partial t} + \left(\frac{\partial V}{\partial x}, f\right) \leq -\nu(\|x\|) \tag{5.67}$$

where $\lambda(r)$, $\mu(r)$, and $\nu(r)$ belong to the class K.

Apparently, there is no comparison equation of the form (5.54). Actually, conditions (5.66) and (5.67) lead easily to a comparison equation. Indeed, from the second inequality (5.66), we get $\|x\| \geq \mu^{-1}(V(t, x))$ which combined with (5.67) yields

$$\frac{\partial V}{\partial t} + \left(\frac{\partial V}{\partial x}, f\right) \leq -\nu(\mu^{-1}(V(t, x))). \tag{5.68}$$

It is useful to point out that the inverse of a function from K also belongs to K and the superposition of two functions from K belongs to K (possibly on $[0, a')$, with $a' < a$).

Let us denote $\varphi(r) = \nu(\mu^{-1}(r))$. It follows that the comparison equation we have to investigate can be written

$$\dot{y} = -\varphi(y) \tag{5.69}$$

our task being now to establish that the solution $y = 0$ of (5.69) is uniformly asymptotically stable. Inasmuch as Eq. (5.69) is with separable variables, we find

$$\Phi(y(t; t_0, y_0)) - \Phi(y_0) = -(t - t_0), \qquad t \geq t_0 \tag{5.70}$$

where $y_0 > 0$, and $\Phi'(u) = 1/\varphi(u)$, $u > 0$. For $t \to \infty$, the right-hand side in (5.70) tends toward $-\infty$. Hence, $\Phi(y(t; t_0, y_0)) \to -\infty$ as $t \to \infty$, which implies $y(t; t_0, y_0) \to 0$ and $\Phi(u) \to -\infty$ as $u \to 0$. Consequently, $\Phi^{-1}(u) \to 0$ as $u \to -\infty$. But (5.70) is equivalent to

$$y(t; t_0, y_0) = \Phi^{-1}(\Phi(y_0) - (t - t_0)). \tag{5.71}$$

From (5.71) we can derive easily the uniform asymptotic stability of the solution $y = 0$ of (5.69), which proves the last corollary.

Remark. We did not discuss in the corollaries the case of asymptotic stability. In the next section, we shall give further results concerning this concept.

Let us establish now a general theorem for *instability*. We take the opportunity to point out that there exist several concepts of instability. In what follows, we shall mean by *instability* the property that is the contrary of the stability property. In other words, the solution $x = 0$ of (5.1) will be termed *unstable* if there exists a number $\varepsilon_0 > 0$ with the following property: There are solutions $x(t; 0, x^0)$ with arbitrarily small $\|x^0\|$, such that $\|x(t; 0, x^0)\| > \varepsilon_0$ for some convenient $t > 0$.

Theorem 5.6

Consider the system (5.1) *under the general assumption from* §5.1. *Assume that there exists a function* $V(t, x)$ *defined on an open set* $G \subset \Delta$, *satisfying on this set the inequality*

$$\frac{\partial V}{\partial t} + \left(\frac{\partial V}{\partial x}, f\right) \geq \omega(t, V) \tag{5.72}$$

and such that the following conditions hold: (1) $0 < V(t, x) \leq M < +\infty$ *on* G; (2) $V(t, x) = 0$ *for* $(t, x) \in \partial G \cap \Delta$; (3) *the half-axis* $x = 0$, $t \geq 0$ *meets* ∂G; (4) $\omega(t, V(t, x)) \geq 0$ *for* $(t, x) \in G$; (5) *the solution* $y(t; t_0, y_0)$ *of the comparison equation* (5.54), *with* y_0 *arbitrarily small, is unbounded on* $[t_0, \infty)$. *Then the solution* $x = 0$ *of the system* (5.1) *is unstable.*

Proof. From condition 3 in Theorem 5.6 follows the existence of some points $(t_0, x^0) \in G$, with arbitrarily small $\|x^0\|$. Condition 4 shows that $V(t, x(t; t_0, x^0))$, with $(t_0, x^0) \in G$, is a nondecreasing function. Hence,

$$V(t, x(t; t_0, x^0)) \geq V(t_0, x^0) = y_0 > 0 \tag{5.73}$$

for all $t \geq t_0$ belonging to the maximal interval of existence of $x(t; t_0, x^0)$. Moreover, (5.73) implies that $(t, x(t; t_0, x^0)) \in G$ on the whole interval of existence of the solution $x(t; t_0, x^0)$, if we consider condition 2.

On the other hand, (5.72) gives

$$V(t, x(t; t_0, x^0)) \geq y(t; t_0, y_0) \tag{5.74}$$

for all $t \geq t_0$, such that $x(t; t_0, x^0)$ is defined. Taking into account condition 5, we get a contradiction from (5.74). Indeed, if $x(t; t_0, x^0)$ is defined on the half-axis $t \geq t_0$, then the right-hand side of (5.74) becomes infinite—

that is, $V(t, x)$ cannot be bounded on G. This fact contradicts condition 1. If the maximal interval of existence for $x(t; t_0, x^0)$ is finite, then there exist values of $t > t_0$, such that $\|x(t; t_0, x^0)\| > a_0$ $(a_0 < a)$, for $\|x^0\|$ arbitrarily small. Hence, $x = 0$ cannot be stable.

Corollary

(*Second Liapunov's theorem of instability.*) *If there exist a function* $V(t, x)$ *and a set* $G \subset \Delta$, *satisfying conditions* 1, 2, *and* 3 *from Theorem* 5.6, *and*

$$\frac{\partial V}{\partial t} + \left(\frac{\partial V}{\partial x}, f\right) \geq \alpha V \tag{5.75}$$

where $\alpha > 0$, *then* $x = 0$ *is unstable for the system* (5.1).

For the proof, it suffices to take $\omega(t, y) = \alpha y$. Then

$$y(t; t_0, y_0) = y_0 \exp[\alpha(t - t_0)]$$

and conditions 4 and 5 are obviously verified.

Another problem we shall consider is that of boundedness of solutions. If all the solutions of the system (5.1) with $f(t, x)$ given in the half-space $t \geq 0$, $x \in R^n$ are bounded, then the system itself is called *bounded.* Very often, this property is called *Lagrange stability.* The comparison method is applicable and the following result holds.

Theorem 5.7

Consider the system (5.1) *with* $f(t, x)$ *continuous and locally Lipschitzian in the half-space* $t \geq 0$, $x \in R^n$. *Assume that there exists a function* $V(t, x)$ *defined on* $t \geq 0$, $x \in R^n$, *such that*: (1) $\lambda(\|x\|) \leq V(t, x)$, *where* $\lambda(r)$ *is a continuous increasing function on* $[0, \infty)$, *with* $\lambda(r) \to \infty$ *as* $r \to \infty$; (2) *inequality* (5.53) *holds*; (3) *the solutions of the comparison equation* (5.54) *are bounded in the future. Then the solutions of* (5.1) *are bounded.*

Proof. It follows easily from the inequality

$$V(t, x(t; t_0, x^0)) \leq y(t; t_0, y_0), \qquad y_0 = V(t_0, x^0) \tag{5.76}$$

which holds on the half-axis $t \geq t_0$.

Indeed, the conditions of Theorem 5.7 are stronger than those of Theorem 3.6, which guarantee the global existence in the future. The boundedness is now the consequence of the inequality we can derive from (5.76)

$$\lambda(\|x(t; t_0, x^0)\|) \le y(t; t_0, y_0), \qquad t \ge t_0. \tag{5.77}$$

Theorem 5.7 is proved.

§5.5. Further Stability Results

We shall consider in this section some stability problems that are not necessarily related to comparison technique. First, we want to give a result concerning asymptotic stability. As we have seen in the preceding section, the existence of a Liapunov function $V(t, x)$—which is positive definite, decrescent, and the derivative of which with respect to the system is negative definite—suffices for the uniform asymptotic stability. On the other hand, the existence of a positive-definite function with negative definite derivative does not imply asymptotic stability (there exist some examples too complicated to be included here). The following theorem gives an answer to the question, what conditions have to be added to the existence of a positive-definite function, with negative-definite derivative, to assure the asymptotic stability of the solution $x = 0$ of (5.1)?

Theorem 5.8

Consider the system (5.1) *under the general assumption from* §5.1. *Assume there exists a positive-definite function* $V(t, x)$ *on* Δ, *the derivative of which with respect to* (5.1) *is negative definite. Moreover, if either* $f(t, x)$ *is bounded on* Δ *or* $x = 0$ *is uniformly stable, then* $x = 0$ *is asymptotically stable for* (5.1).

Proof. Considering that our conditions imply the stability of the solution $x = 0$, we have to prove that

$$\lim_{t\to\infty} \|x(t; t_0, x^0)\| = 0 \tag{5.78}$$

for all x^0 such that $\|x^0\| < \gamma(t_0)$.

Let us assume first that $f(t, x)$ is bounded on Δ (or even on $t \ge 0$, $\|x\| \le a' < a$). Suppose that $x = 0$ is not asymptotically stable. Then there exist a solution $x(t; t_0, x^0)$ of (5.1) and a sequence $\{t_k\}$, $t_k \to \infty$, such that $\|x(t_k; t_0, x^0)\| \ge \varepsilon_0 > 0$, $k = 1, 2, \ldots$. Because $\|f(t, x)\| \le M$, we can construct the intervals $[t_k - \varepsilon_0(2M)^{-1}, t_k + \varepsilon_0(2M)^{-1}]$, $k = 1, 2, \ldots$, and, moreover, we can assume them to be disjoint and $t_1 > t_0 + \varepsilon_0(2M)^{-1}$. If necessary, one can replace $\{t_k\}$ by a convenient subsequence.

For $t \in [t_k - \varepsilon_0(2M)^{-1}, t_k + \varepsilon_0(2M)^{-1}]$, we have $\|x(t; t_0, x^0)\| \geq \varepsilon_0/2$. Indeed, from $\|f(t, x)\| \leq M$, we obtain $\|x(t; t_0, x^0) - x(s; t_0, x^0)\| \leq M|t - s|$. Taking into account that $\|x(t_k; t_0, x^0)\| \geq \varepsilon_0$, $k = 1, 2, \ldots$, we get $\|x(t; t_0, x^0)\| \geq \varepsilon_0/2$ for $|t - t_k| \leq \varepsilon_0(2M)^{-1}$.

Let $\lambda(r)$ and $\mu(r)$ be some functions of class K such that $\lambda(\|x\|) \leq V(t, x)$ and

$$\frac{\partial V}{\partial t} + \left(\frac{\partial V}{\partial x}, f\right) \leq -\mu(\|x\|). \tag{5.79}$$

This means that for $x = x(t; t_0, x^0)$, $|t - t_k| \leq \varepsilon_0(2M)^{-1}$, we have

$$\frac{\partial V}{\partial t} + \left(\frac{\partial V}{\partial x}, f\right) \leq -\mu\left(\frac{\varepsilon_0}{2}\right). \tag{5.80}$$

From (5.80), we obtain by integration

$$V(t_k + \varepsilon_0(2M)^{-1}, x(t_k + \varepsilon_0(2M)^{-1}; t_0, x^0)) - V(t_k - \varepsilon_0(2M)^{-1}, x(t_k - \varepsilon_0(2M)^{-1}; t_0, x^0)) \leq -\varepsilon_0 M^{-1}\mu(\varepsilon_0/2). \tag{5.81}$$

But (5.79) shows that $V(t, x(t; t_0, x^0))$ is a decreasing function on $t \geq t_0$. Hence, (5.81) yields

$$V(t_k + \varepsilon_0(2M)^{-1}, x(t_k + \varepsilon_0(2M)^{-1}; t_0, x^0)) - V(t_0, x^0) < -k\varepsilon_0 M^{-1}\mu(\varepsilon_0/2). \tag{5.82}$$

Let us make $k \to \infty$ in (5.82). The right-hand side tends toward $-\infty$, which contradicts $V(t, x) \geq 0$. This shows the asymptotic stability of the solution $x = 0$ of (5.1).

Assume now, instead of boundedness of $f(t, x)$, the uniform stability of $x = 0$. Of course, we have to prove (5.78) only. It is interesting to point out that, under our assumptions, (5.78) holds uniformly in x^0, $\|x^0\| < \delta_0$.

Inasmuch as $x = 0$ is uniformly stable, given any $\varepsilon > 0$, $\varepsilon \leq a' < a$, there exists $\delta(\varepsilon) > 0$ such that $\|x(t; t_0, x^0)\| < \varepsilon$ for $t \geq t_0$ and $\|x^0\| < \delta(\varepsilon)$. Let us denote $\delta_0 = \delta(a')$. We shall prove that given any $t_0 \geq 0$ and x^0, with $\|x^0\| < \delta_0$, there exists $\tau(\varepsilon, t_0) > 0$ such that $\|x(s; t_0, x^0)\| < \delta(\varepsilon)$ for some $s \in [t_0, t_0 + \tau]$. This property implies $\|x(t; t_0, x^0)\| < \varepsilon$ for $t \geq s$ and, a fortiori, for $t \geq t_0 + \tau(\varepsilon, t_0)$—that is, the asymptotic stability of the solution $x = 0$ of (5.1).

Let us prove now the existence of s in the preceding assertion. Given any $\varepsilon > 0$, $\varepsilon < a'$, from $t \geq 0$, $\delta(\varepsilon) \leq \|x\| < a'$, it follows that $V(t, x) \geq \tilde{\lambda}(\varepsilon) = \lambda(\delta(\varepsilon))$ and

$$\frac{\partial V}{\partial t} + \left(\frac{\partial V}{\partial x}, f\right) \leq -\tilde{\mu}(\varepsilon). \tag{5.83}$$

Let $v(t_0) = \sup\{V(t_0, x); \|x\| < \delta_0\}$ and denote $\tau(\varepsilon, t_0) = v(t_0)/\tilde{\mu}(\varepsilon)$. Then, given x^0 with $\|x^0\| < \delta_0$, either $\|x^0\| \geq \delta(\varepsilon)$ or not. In the first case, $\|x(t; t_0, x^0)\| \geq \delta(\varepsilon)$ for some $t \geq t_0$. If $\delta(\varepsilon) \leq \|x(t; t_0, x^0)\| < a'$ on $[t_0, t_0 + \tau]$, from (5.83) we get

$$\tilde{\lambda}(\varepsilon) \leq V(t_0 + \tau, x(t_0 + \tau; t_0, x^0)) \leq V(t_0, x^0) - \tau\tilde{\mu}(\varepsilon) \leq 0$$

which contradicts the fact that $\tilde{\lambda}(r) > 0$ for $r > 0$. Hence, there is at least one $s \in [t_0, t_0 + \tau]$ such that $\|x(s; t_0, x^0)\| < \delta(\varepsilon)$, which ends the proof.

Remark. Theorem 5.8 is particularly useful when $f(t, x) \equiv f(x)$—that is, (5.1) is an autonomous system—or $f(t + \omega, x) \equiv f(t, x)$ for a given $\omega > 0$ (the periodic case).

We shall now make a few comments concerning the *asymptotic stability in the large*. This concept will be used in the next section, when we shall consider the stability problem of some automatic control systems.

Consider the system (5.1) with $f(t, x)$ given on $[0, \infty) \times R^n$ (i.e., $a = \infty$ in the definition of Δ), under the same general assumptions on $f(t, x)$. The solution $x = 0$ of (5.1) is called *asymptotically stable in the large* if it is stable, and $\lim\|x(t; t_0, x^0)\| = 0$ as $t \to \infty$, for any $t_0 \geq 0$, $x^0 \in R^n$.

From Theorem 5.8, the following criterion of asymptotic stability in the large can be easily derived.

Theorem 5.9

Consider system (5.1) *with $f(t, x)$ continuous and locally Lipschitzian on $[0, \infty) \times R^n$, bounded on any half-cylinder $t \geq 0$, $\|x\| \leq R < +\infty$, and such that $f(t, 0) \equiv 0$. Moreover, assume there exists a Liapunov function $V(t, x)$ on $[0, \infty) \times R^n$, positive definite*

$$\lambda(\|x\|) \leq V(t, x) \tag{5.84}$$

and with a negative-definite derivative

$$\frac{\partial V}{\partial t} + \left(\frac{\partial V}{\partial x}, f\right) \leq -\mu(\|x\|). \tag{5.85}$$

If $\lambda(r) \to \infty$ as $r \to \infty$, then $x = 0$ is asymptotically stable in the large.

Let us remark that our conditions imply those of both Theorem 5.7 and 5.8 (for the latter, with respect to any half-cylinder $t \geq 0$, $\|x\| \leq R < +\infty$). This leads immediately to the statement of Theorem 5.9.

Another problem we want to discuss in this section is that of existence (or construction) of Liapunov functions. When $x = 0$ possesses a certain stability property with respect to the system (5.1), it is natural to ask whether a Liapunov function $V(t, x)$ exists or not. A result of this kind is often called a *converse theorem* in stability theory. We shall restrict our considerations to the case of linear systems of the form

$$\dot{x} = A(t)x \tag{5.15}$$

which can be treated relatively easily. The case of nonlinear systems requires a more intricate discussion, and we are not going to consider it in this book.

Theorem 5.10

Consider the system (5.15), *where* $A(t)$ *is a continuous bounded matrix on* $t \geq 0$, *such that* $x = 0$ *is uniformly asymptotically stable. Then, given any constant square matrix* C, *such that* $C = C^*$ *and* (Cx, x) *is positive definite, there exists* $B(t) = B^*(t)$ *satisfying*

$$\dot{B} + BA + A^*B = -C, \qquad t \geq 0 \tag{5.86}$$

with the property that the quadratic form

$$V(t, x) = (B(t)x, x) \tag{5.87}$$

is positive definite, decrescent, and its derivative with respect to (5.15) *is* $-(Cx, x)$.

Proof. First, let us remark that the derivative of $V(t, x)$ given by (5.87), with respect to (5.15), is

$$\frac{\partial V}{\partial t} + \left(\frac{\partial V}{\partial x}, A(t)x\right) = ((\dot{B} + BA + A^*B)x, x). \tag{5.88}$$

This equation can be obtained by direct calculation, taking into account $(Ax, y) = (x, A^*y)$, a relation that characterizes the adjoint operators (matrices). From (5.86) and (5.88), it follows that the derivative of $V(t, x)$ is $-(Cx, x)$.

We shall prove that

$$B(t) = \int_t^\infty [X(s)X^{-1}(t)]^*C[X(s)X^{-1}(t)]\, ds \tag{5.89}$$

where $X(t)$ is the fundamental matrix of (5.15) with $X(0) = I$, satisfies the required properties. That (5.89) has a meaning follows from the estimate

$$\|X(s)X^{-1}(t)\| \leq Ne^{-\alpha(s-t)}, \qquad s \geq t \geq 0 \tag{5.90}$$

(see Theorem 5.2), which is a consequence of the uniform asymptotic stability. To prove (5.86), we can proceed by direct calculation, considering that

$$\frac{d}{dt} X^{-1}(t) = -X^{-1}(t)A(t), \qquad \frac{d}{dt}(X^{-1}(t))^* = -A^*(t)(X^{-1}(t))^*.$$

For instance, the first relation can be obtained from $X(t)X^{-1}(t) = I$, by differentiation.

From (5.87) and (5.89), we get

$$V(t, x) = \int_t^\infty (CX(s)X^{-1}(t)x, X(s)X^{-1}(t)x)\, ds, \qquad t \geq 0, \qquad x \in R^n. \tag{5.91}$$

Inasmuch as

$$V(t, x) \leq \|C\| \cdot \|x\|^2 \int_t^\infty \|X(s)X^{-1}(t)\|^2\, ds$$

we find, taking into account (5.90),

$$V(t, x) \leq \frac{N^2 \|C\|}{2\alpha} \|x\|^2$$

which shows that $V(t, x)$ is decrescent.

It remains to prove that $V(t, x)$ is positive definite. From $\dot{X} = AX$ we obtain

$$X(s) = X(t) + \int_t^s A(u)X(u)\, du$$

which yields

$$X(s)X^{-1}(t) = I + \int_t^s A(u)X(u)X^{-1}(t)\, du = I + \Phi(s, t). \tag{5.92}$$

Inasmuch as $\|A(t)\| \leq A_0$ on $t \geq 0$, it follows from (5.90) that

$$\|\Phi(s, t)\| \leq \frac{A_0 N}{\alpha} [1 - e^{-\alpha(s-t)}], \qquad s \geq t \geq 0$$

from which we obtain

$$\|\Phi(s, t)\| < \varepsilon \quad \text{for} \quad 0 \leq t \leq s \leq t + T(\varepsilon). \tag{5.93}$$

Taking into account (5.91), (5.92), and (5.93), we find

$$V(t, x) \geq \int_t^{t+T} (Cx + C\Phi x, x + \Phi x)\, ds \geq T\lambda \|x + \Phi x\|^2$$

where λ is the smallest characteristic root of C. Hence, for sufficiently small ε, we have $\|x+\Phi x\| \geq (1-\varepsilon)\|x\|$ and

$$V(t, x) \geq \gamma\|x\|^2, \qquad \gamma = T\lambda(1-\varepsilon)^2 \tag{5.94}$$

which ends the proof of Theorem 5.10.

Corollary

Let A be a square matrix with negative characteristic roots. Then there exists a square matrix $B = B^*$, such that (Bx, x) is positive definite and $BA + A^*B = -C$, where C is a given symmetric matrix for which (Cx, x) is a positive definite quadratic form.

The corollary corresponds to the case $A(t) = A = \text{const}$, when B can be chosen also to be a constant matrix. An application of the corollary will be given in the next section.

§5.6. Stability of Automatic Control Systems

The general theorems on stability find many important applications in various fields. One of the most interesting problems that can be approached through use of Liapunov functions is concerned with the theory of automatic control. More precisely, we shall consider a problem formulated first by Lur'e and Postnikov, related to the stability of some systems occurring in the theory of automatic control.

Let us consider the system

$$\dot{x} = Ax + bf(\sigma), \qquad \sigma = (c, x) \tag{5.95}$$

where x is an n-vector, A is a square matrix with negative characteristic roots, b and c are constant n-vectors, and $f(\sigma)$ is a real-valued function defined on the whole real axis. The function $f(\sigma)$ is generally nonlinear and is called the *characteristic of the servomotor*. It is only approximately known, a circumstance which makes desirable that the stability properties of the system be independent of the special choice of $f(\sigma)$, within a given class.

Throughout this section, it will be assumed that $f(\sigma)$ is a continuous function on the real axis, such that

$$\sigma f(\sigma) > 0, \qquad \sigma \neq 0. \tag{5.96}$$

This class of functions will be denoted by $\mathscr{C}$. It is obvious that (5.96) and the continuity of $f(\sigma)$ imply $f(0) = 0$.

We shall give now the definition of the concept of *absolute stability*. System (5.95) is called *absolutely stable* if $x = 0$ is asymptotically stable in the large, no matter how we choose the function $f(\sigma) \in \mathscr{C}$. Inasmuch as $f(\sigma)$ does not necessarily satisfy a local Lipschitz condition, the uniqueness property is not always true. Of course, we can assume it holds, although our considerations remain valid without this assumption. Further arguments would be necessary in the last case, but we are not going to give more details here.

To apply the criterion given by Theorem 5.9 for global asymptotic stability, it is convenient to put in another form the system (5.95). Namely, if we differentiate the equation $\sigma = (c, x)$ and take into account the first equation of (5.95), we find

$$\dot{\sigma} = (c, \dot{x}) = (c, Ax) + (c, b)f(\sigma).$$

Hence, the following system can be associated with (5.95):

$$\dot{x} = Ax + bf(\sigma), \qquad \dot{\sigma} = (c, Ax) + (c, b)f(\sigma). \tag{5.97}$$

System (5.97) is an autonomous system in $x_1, x_2, \ldots, x_n, \sigma$ and if the zero solution of (5.97) is asymptotically stable in the large, the same property holds for the zero solution of the system (5.95). Inasmuch as $(c, Ax) = (A^*c, x)$, we can denote A^*c again by c, and setting $r = -(c, b)$, where c is the initial vector occurring in our system, (5.97) becomes

$$\dot{x} = Ax + bf(\sigma), \qquad \dot{\sigma} = (c, x) - rf(\sigma). \tag{5.98}$$

We shall formulate the stability result for the system (5.98), leaving to the reader the task of restating it for the system (5.97). First, consider a symmetric matrix $C = C^*$ with (Cx, x) positive definite, and let $B = B^*$ be the matrix satisfying $BA + A^*B = -C$ (see the corollary of Theorem 5.10). Then (Bx, x) is positive definite.

For the system (5.98), consider the Liapunov function

$$V(x, \sigma) = (Bx, x) + \int_0^\sigma f(u)\, du. \tag{5.99}$$

Let us denote $F(\sigma) = \int_0^\sigma f(u)\, du$. Then $F(\sigma) > 0$ for $\sigma \neq 0$ and $F(0) = 0$. Inasmuch as $\lambda\|x\|^2 \leq (Bx, x) \leq \Lambda\|x\|^2$, with $\lambda > 0$, it follows that

$$\lambda\|x\|^2 + F(\sigma) \leq V(x, \sigma) \leq \Lambda\|x\|^2 + F(\sigma). \tag{5.100}$$

This shows that $V(x, \sigma)$ is positive definite and decrescent. Moreover, if we assume

$$\lim F(\sigma) = +\infty \quad \text{as} \quad |\sigma| \to +\infty \tag{5.101}$$

we find that $\|x\| + |\sigma| \to \infty$ implies $V(x, \sigma) \to \infty$, as required in Theorem 5.9.

We have only to prove that the derivative of $V(x, \sigma)$ is negative definite. More precisely, we are going to find appropriate conditions to ensure this last property. From (5.99) we get

$$\begin{aligned}\frac{d}{dt} V(x, \sigma) &= (B\dot{x}, x) + (Bx, \dot{x}) + f(\sigma)\dot{\sigma} \\ &= (BAx, x) + (Bx, Ax) + (Bb, x)f(\sigma) + (Bx, b)f(\sigma) \\ &\quad + ((c, x) - r(f(\sigma))f(\sigma) = ((BA + A^*B)x, x) \\ &\quad + 2(Bb, x)f(\sigma) + (c, x)f(\sigma) - rf^2(\sigma) \\ &= -(Cx, x) + 2(Bb + \tfrac{1}{2}c, x)f(\sigma) - rf^2(\sigma)\end{aligned}$$

which shows that the derivative of V is a quadratic form in $x_1, x_2, \ldots, x_n$ and $f(\sigma)$.

This feature is the key to the method we are developing, because the problem is now reduced to a well-known algebraic criterion. We have in mind the Sylvester procedure.[2]

The condition we want to ensure is that

$$(Cx, x) - 2(Bb + \tfrac{1}{2}c, x)f(\sigma) + rf^2(\sigma) \tag{5.102}$$

is a positive definite quadratic form. The symmetric matrix associated with (5.102) is

$$D = \begin{pmatrix} C & -(Bb + \frac{1}{2}c) \\ -(Bb + \frac{1}{2}c)^* & r \end{pmatrix}. \tag{5.103}$$

Inasmuch as all the main determinants of order less than or equal to n are positive [the matrix C is by hypothesis such that (Cx, x) is positive definite], the only condition we have to impose is $\det D > 0$. Because

$$\det\begin{pmatrix} C^{-1} & 0 \\ 0 & 1 \end{pmatrix} > 0$$

$\det D > 0$ is equivalent to

$$\det\begin{pmatrix} C^{-1} & 0 \\ 0 & 1 \end{pmatrix} D > 0.$$

A simple calculation shows that the product of the matrices

$$\begin{pmatrix} C^{-1} & 0 \\ 0 & 1 \end{pmatrix}$$

and D is the matrix

$$\begin{pmatrix} I & -C^{-1}(Bb + \frac{1}{2}c) \\ -(Bb + \frac{1}{2}c)^* & r \end{pmatrix}$$

[2] For instance, I. M. Gel'fand, *Lectures on Linear Algebra*, Interscience Publishers, New York, 1961.

which finally yields the following condition:

$$r > (Bb + \tfrac{1}{2}c)^* C^{-1}(Bb + \tfrac{1}{2}c). \tag{5.104}$$

An equivalent form of inequality (5.104) is

$$r > (C^{-1}(Bb + \tfrac{1}{2}c), (Bb + \tfrac{1}{2}c)). \tag{5.105}$$

In particular, (5.104) implies $r > 0$, because C^{-1} generates also a positive definite quadratic form. Indeed, from $(Cx, x) > 0$ for $x \neq 0$, we obtain $(y, C^{-1}y) > 0$ for $y \neq 0$, where $Cx = y$.

Consequently, *system* (5.98) *is absolutely stable if A has all its characteristic roots with negative real parts, and, moreover, given $C = C^*$ with (Cx, x) positive definite, condition* (5.105) *holds, where $B = B^*$ is the solution of $BA + A^*B = -C$, such that (Bx, x) is also positive definite.*

Of course, an interesting optimal problem arising in this approach to the stability of automatic control systems is to minimize the right-hand side of (5.104) with respect to C. Condition (5.104) does not involve anything connected to the nonlinear part of the system [i.e., to $f(\sigma)$], as required by the application needs.

PROBLEMS

1. Consider the second-order linear differential equation $\ddot{x} + \omega^2 x = f(t)$, with $\omega \neq 0$ and $f(t)$ continuous on $t \geq 0$, such that $\int_0^\infty |f(t)|\, dt < \infty$. Prove that any solution is bounded on $t \geq 0$.
2. If ω and $f(t)$ satisfy the assumptions from problem 1, then the solution $x = 0$ of the equation $\ddot{x} + [\omega^2 + f(t)]x = 0$ is uniformly stable.
3. Use Ważewski's inequality [problem 4, Chapter 4] to find conditions that ensure the stability (asymptotic stability) of the solution $x = 0$ of the system $x = A(t)x$, where $A(t)$ is a continuous matrix on $t \geq 0$.
4. Consider the linear system $\dot{x} = A(t)x + f(t)$, with $A(t)$ and $f(t)$ continuous on $t \geq 0$. Assume that the solution $x = 0$ of the homogeneous system $\dot{x} = A(t)x$ is uniformly stable and $f(t)$ is absolutely integrable on $[0, \infty)$. Show that any solution of the nonhomogeneous system is bounded on $t \geq 0$.
5. Let A be a constant n by n matrix with characteristic roots lying in the half-plane $\mathrm{Re}(\lambda) < 0$. If $B(t)$ is a continuous n by n matrix such that $\|B(t)\| \to 0$ as $t \to \infty$, then the solution $x = 0$ of the system $\dot{x} = [A + B(t)]x$ is asymptotically stable.
   Hint: Use the equivalent integral equation
   $$x(t) = \exp[A(t - t_0)]x(t_0) + \int_{t_0}^{t} e^{A(t-s)}B(s)x(s)\, ds.$$
6. Consider the linear differential system with polynomial coefficients
   $$\dot{x} = (A_0 t^m + A_1 t^{m-1} + \cdots + A_m)x$$
   where A_k, $(k = 0, 1, \ldots, m)$ are constant n by n matrices. Assume that the

matrix A_0 has its characteristic roots with negative real parts. Prove that the solution $x=0$ is asymptotically stable.
Hint: Introduce a new independent variable $s=(m+1)^{-1}t^{m+1}$ and apply the result from problem 5.

7. Consider the differential system $\dot{x}=A(t)x+f(t)$, with continuous $A(t)$ and $f(t)$ on the positive half-axis. Assume that the solution $x=0$ of the associated homogeneous system is uniformly asymptotically stable and $\|f(t)\|\to 0$ as $t\to\infty$. Prove that any solution of the given system tends toward zero as $t\to\infty$.

8. Let $f(t)$ be a continuous nonnegative function such that

$$\int_t^{t+1} f(s)\,ds \leq C, \qquad t \geq 0.$$

Then, for any $\alpha>0$,

$$e^{-\alpha t}\int_0^t e^{\alpha s}f(s)\,ds \leq (1-e^{-\alpha})^{-1}C, \qquad t\geq 0.$$

9. Consider the differential system $\dot{x}=A(t)x+f(t)$, with $A(t)$ and $f(t)$ continuous on $t\geq 0$. Moreover, assume that the solution $x=0$ of the corresponding homogeneous system is uniformly asymptotically stable (hence, exponentially asymptotically stable) and $f(t)$ is such that

$$\int_t^{t+1} \|f(s)\|\,ds \leq C, \qquad t\geq 0.$$

Show that any solution of the nonhomogeneous system is bounded.
Hint: Use the formula of variation of parameters and the result stated in problem 8.

10. Consider the system $\dot{x}=A(t)x+f(t,x)$, with $A(t)$ continuous on $t\geq 0$ and $f(t,x)$ continuous and locally Lipschitzian for $t\geq 0$, $\|x\|<a\leq\infty$. Assume further that $\|f(t,x)\|\leq\lambda(t)\|x\|$. If the solution $x=0$ of the system $\dot{x}=A(t)x$ is uniformly asymptotically stable, find conditions on $\lambda(t)$ such that the zero solution of the perturbed system is asymptotically stable.

11. Investigate the stability properties of the system $\dot{x}=[A(t)+B(t)]x+f(t,x)$, under suitable conditions concerning the matrices $A(t)$, $B(t)$ and the vector function $f(t,x)$. Assume that $f(t,0)\equiv 0$ and that $f(t,x)$ is "small" in some sense. In particular, consider the case when $A(t)$ reduces to a constant matrix.

12. Consider the second-order scalar equation $\ddot{x}=r(x,\dot{x})$, with $r(x,y)$ locally Lipschitzian in a neighborhood of the origin $(0,0)$. Assume further that $r(0,0)=0$. Notice that the equation is equivalent to the system $\dot{x}=y$, $\dot{y}=r(x,y)$, and investigate the boundedness and stability properties using the Liapunov function

$$V(x,y)=y^2-2\int_0^x r(u,0)\,du.$$

13. The equations describing the motion of a rigid body (gyroscope) that rotates about a fixed point are $A\dot{p}=(B-C)qr$, $B\dot{q}=(C-A)rp$, $C\dot{r}=(A-B)pq$.

Constants A, B, and C denote the main moments of inertia, and p, q, and r the coordinates of the velocity vector with respect to the principal axes of inertia. It is obvious that the system possesses the following equilibrium states: $(p_0, 0, 0)$, $(0, q_0, 0)$, and $(0, 0, r_0)$. Here, p_0, q_0, and r_0 are arbitrary constants. In studying the stability of the state $(0, 0, r_0)$, $r_0 > 0$, let us introduce the new variables $u = p$, $v = q$, $w = r - r_0$. Assume further that $A \geq B > C$, and investigate the stability using the Liapunov function

$$V = \frac{A-C}{B}u^2 + \frac{B-C}{A}v^2 + [Au^2 + Bv^2 + 2Cr_0w + Cw^2]^2.$$

14. Consider the system $\dot{x} = a(t)y + b(t)x(x^2 + y^2)$, $\dot{y} = -a(t)x + b(t)y(x^2 + y^2)$, with $a(t)$ and $b(t)$ continuous on $t \geq 0$. Investigate the stability properties using the function $V(x, y) = x^2 + y^2$.

15. Let us consider the second-order equation $\ddot{x} + p(t)\dot{x} + q(t)f(x) = 0$, where $p(t)$ is continuous on $t \geq 0$ and $q(t)$ is continuously differentiable on $t \geq 0$. Assume that $0 < q(t) \leq M$, $2p(t)q(t) + \dot{q}(t) \geq 0$, and $\int_0^{\pm\infty} f(x)\,dx = +\infty$. Prove that any solution of the given equation is bounded on $t \geq 0$, together with its first derivative.
Hint: Transform the equation into an equivalent system and use the Liapunov function

$$V(t, x, y) = \int_0^x f(u)\,du + \frac{y^2}{2q(t)}.$$

16. Consider the differential system $\dot{x} = f(t, x)$ with $f(t, x)$ continuous and locally Lipschitzian in x for $t \geq 0$, $\|x\| < a \leq \infty$. Assume further that $f(t, 0) \equiv 0$ and there exists a positive-definite Liapunov function $V(t, x)$, such that

$$\frac{\partial V}{\partial t} + \left(\frac{\partial V}{\partial x}, f\right) \leq -\varphi(t)\lambda(V).$$

If $\varphi(t)$ is a continuous function on $t \geq 0$ such that $\int_0^\infty \varphi(t)\,dt = \infty$, and $\lambda(r)$ satisfies the usual conditions from §5.4, then the solution $x = 0$ of the system $\dot{x} = f(t, x)$ is asymptotically stable.
Hint: Use the comparison equation $\dot{y} = -\varphi(t)\lambda(y)$.

17. Consider the system $\dot{x} = f(t, x)$, $f(t, 0) \equiv 0$, and assume that

$$(x, f(t, x)) \leq \|x\|\lambda(\|x\|)g(t)$$

with $g(t)$ continuous on $t \geq 0$ and $\lambda(r)$ such that

$$\int_0^{r_0 > 0} \frac{dr}{\lambda(r)} = \infty.$$

Investigate the stability properties of the solution $x = 0$ of the system $\dot{x} = f(t, x)$.
Hint: Find a Liapunov function of the form $V(t, x) = \alpha(\|x\|)$, such that the following inequality is valid:

$$\frac{\partial V}{\partial t} + \left(\frac{\partial V}{\partial x}, f\right) \leq g(t)V.$$

18. Consider the differential system $\dot{x} = f(t, x)$, $f(t, 0) \equiv 0$, with $f(t, x)$ continuous and locally Lipschitzian on $t \geq 0$, $\|x\| < a \leq \infty$. Assume that there exists a Liapunov function $V(t, x)$ satisfying conditions (5.66) and (5.67). Then there exists a Liapunov function $W(t, x)$ that is positive definite, decrescent, and the derivative of which, with respect to the system, satisfies

$$\frac{\partial W}{\partial t} + \left(\frac{\partial W}{\partial x}, f\right) \leq -W(t, x).$$

Hint: Find $W(t, x)$ of the form $W(t, x) = \lambda(V(t, x))$, $\lambda \in K$.

19. Let $\dot{x} = f(t, x)$ be a differential system satisfying the conditions listed in problem 18. Assume further that $V(t, x)$ satisfies a Lipschitz condition: $|V(t, x) - V(t, y)| \leq L\|x - y\|$. Consider the perturbed system $\dot{x} = f(t, x) + R(t, x)$ and find conditions that guarantee the uniform asymptotic stability of the solution $x = 0$.
Remark. The assumptions made on $f(t, x)$ and $V(t, x)$ assure the uniform asymptotic stability of the solution $x = 0$ of the system $\dot{x} = f(t, x)$. Problem 19 consists in finding conditions under which this stability property is preserved for the perturbed system. To solve this problem, use the function $W(t, x)$ constructed as shown in problem 18.

20. Consider the differential system $\dot{x} = f(t, x)$ under usual conditions and show that the exponential asymptotic stability is equivalent to the existence of a Liapunov function $V(t, x)$, satisfying the following conditions: $A\|x\|^2 \leq V(t, x) \leq B\|x\|^2$, $(\partial V/\partial t) + ((\partial V/\partial x), f) \leq -C\|x\|^2$, where A, B, and C are positive constants.
Hint: The sufficiency of these conditions can be proved easily. To prove the necessity, assume that $f(t, x)$ is continuously differentiable in x, and consider

$$V(t, x) = \int_t^{t+\tau} \|x(s; t, x)\|^2 \, ds$$

where $x(t; t_0, x^0)$ has the usual meaning and $\tau = \alpha^{-1} \log(N\sqrt{2})$, the numbers α and N being such that

$$\|x(t; t_0, x^0)\| \leq N\|x^0\| \exp\{-\alpha(t - t_0)\}, \qquad t \geq t_0.$$

21. Consider the differential system $\dot{x} = f(t, x)$, $f(t, 0) \equiv 0$, with $f(t, x)$ continuous on $t \geq 0$, $\|x\| < a \leq \infty$. Assume further that $f(t, x)$ is continuously differentiable with respect to x and, denoting by J the Jacobi matrix $\partial f/\partial x$, that there exists a real symmetric positive-definite matrix B, such that any characteristic root of the symmetric matrix $\frac{1}{2}(J^*B + BJ)$ satisfies $\lambda(t, x) \leq -\delta < 0$, with δ a constant. Then the solution $x = 0$ of the given system is uniformly asymptotically stable.
Hint: Use the Liapunov function $V(t, x) = (Bx, x)$, and show that its derivative is negative definite.

22. Let us consider the system $\dot{x} = xh(x) + ay$, $\dot{y} = bx + cy$, with constant a, b, c and continuous $h(x)$ for all real x. Assume that $h(x) + c < 0$ and $h(x)c - ab > 0$. Moreover, let $h(x)$ be such that $h(x)c - ab > \varepsilon$ for sufficiently large x. Investigate the stability properties of the solution $x = y = 0$.
Hint: Use the Liapunov function

$$V(x, y) = \tfrac{1}{2}(cx - ay)^2 + \int_0^x [h(x)c - ab]x \, dx.$$

23. Consider the system $\dot{x} = Ax + bf(\sigma)$, $\dot{\sigma} = (c, x) - rf(\sigma)$, and assume that there exists a positive-definite matrix H such that $rG - gg^*$ is positive definite, where $-G = HA + A^*H$, $-g = Hb + \frac{1}{2}c$. If the characteristic roots of A have negative real parts and $r + (A^{-1}b, c) > 0$, then the zero solution of the system considered previously is absolutely stable.
Hint: Use the Liapunov function

$$V(x, \sigma) = (Hx, x) + \int_0^\sigma f(u)\, du + \frac{p}{2(r + c^*A^{-1}b)}(c^*A^{-1}x - \sigma)^2$$

with $p > 0$ sufficiently small.

6

Volterra Integral Equations

We have already considered some special Volterra integral equations (scalar or vector)—for example, Eq. (2.3) from §2.1 or Eq. (5.40) from §5.3. These equations were obtained by simple transformations from certain differential equations or systems to which they are equivalent. Now we shall prove that the methods used in studying those particular integral equations can be successfully applied in more general cases.

There are several common features for ordinary differential equations and Volterra integral equations. We shall attempt to emphasize the parallelism between these two theories and to present some main properties pertaining to the latter one.

§6.1. Existence and Uniqueness of Solution

The general (scalar) integral equation of Volterra can be written in the form

$$x(t) = f(t) + \int_{t_0}^{t} k(t, s, x(s))\, ds, \qquad t \in [t_0, t_0 + a]. \tag{6.1}$$

Equation (6.1) is commonly called a Volterra integral equation of the *second kind.* Later on, we shall also consider the Volterra integral equation of the *first kind.*

Theorem 6.1

Let us consider Eq. (6.1) *under the following assumptions*:

1. $f(t)$ *is a continuous (real) function on* $[t_0, t_0 + a]$, $a > 0$;
2. $k(t, s, x)$ *is a continuous function on*

$$\Delta: \qquad t_0 \le s \le t \le t_0 + a, \qquad |x - f(t)| \le b, \qquad b > 0;$$

3. $k(t, s, x)$ *satisfies in* Δ *the Lipschitz condition*

$$|k(t, s, x) - k(t, s, y)| \le L\,|x - y|.$$

Then, there exists a unique continuous solution $x(t)$ *of* (6.1), *defined for*

$$t \in [t_0, t_0 + \delta], \qquad \delta = \min\left\{a, \frac{b}{M}\right\} \tag{6.2}$$

with $M = \sup |k(t, s, x)|$, $(t, s, x) \in \Delta$.

Proof. We shall apply the method of successive approximations, constructing the sequence $\{x_n(t)\}$ as follows:

$$x_0(t) = f(t) \tag{6.3}$$

$$x_n(t) = f(t) + \int_{t_0}^{t} k(t, s, x_{n-1}(s))\, ds, \qquad n \ge 1. \tag{6.4}$$

By mathematical induction one can see that all $x_n(t)$ are continuous functions on $[t_0, t_0 + \delta]$, with δ given by (6.2).

To prove the uniform convergence of the sequence $\{x_n(t)\}$, we shall consider the associated series

$$\sum_{n=1}^{\infty} [x_n(t) - x_{n-1}(t)]. \tag{6.5}$$

From (6.4), for $n = 1$, we get

$$|x_1(t) - x_0(t)| \le M(t - t_0), \qquad t \in [t_0, t_0 + \delta]. \tag{6.6}$$

To obtain the estimates for $|x_n(t) - x_{n-1}(t)|$, $n \ge 2$, we remark that (6.4) gives for $n \ge 2$

$$|x_n(t) - x_{n-1}(t)| \le L \int_{t_0}^{t} |x_{n-1}(s) - x_{n-2}(s)|\, ds, \qquad t \in [t_0, t_0 + \delta] \tag{6.7}$$

taking into account condition 3. From (6.6) and (6.7) we get for $n = 2$

$$|x_2(t) - x_1(t)| \le ML \frac{(t - t_0)^2}{2!}, \qquad t \in [t_0, t_0 + \delta]. \tag{6.8}$$

Inequalities (6.6) and (6.8) suggest the following general estimate for the terms of series (6.5):

$$|x_n(t) - x_{n-1}(t)| \leq ML^{n-1}\frac{(t-t_0)^n}{n!}, \qquad t \in [t_0, t_0 + \delta]. \tag{6.9}$$

If we take into account (6.7), it follows by mathematical induction that (6.9) holds for any n, $n \geq 1$. Consequently, (6.5) is uniformly convergent on $[t_0, t_0 + \delta]$, and so is the sequence $\{x_n(t)\}$.

Denote $x(t) = \lim x_n(t)$ as $n \to \infty$. The function $x(t)$ is continuous on $[t_0, t_0 + \delta]$, and it can be seen that it satisfies (6.1). Indeed, in (6.4) we can make $n \to \infty$ and obtain (6.1), because

$$\lim_{n\to\infty} \int_{t_0}^{t} k(t, s, x_{n-1}(s))\, ds = \int_{t_0}^{t} k(t, s, x(s))\, ds, \qquad t \in [t_0, t_0 + \delta].$$

The last relation holds because

$$|k(t, s, x_{n-1}(s)) - k(t, s, x(s))| \leq L\,|x_{n-1}(s) - x(s)|, \quad t_0 \leq s \leq t \leq t_0 + \delta.$$

Thus, the existence of a solution is proved.

To prove the uniqueness, we shall follow the same method as that in the proof of Theorem 2.1. Let $y(t)$ be a continuous solution of (6.1) on $[t_0, t_0 + \delta]$. Then

$$y(t) = f(t) + \int_{t_0}^{t} k(t, s, y(s))\, ds.$$

Subtracting (6.4) from this last equation, we obtain for $n \geq 1$

$$|y(t) - x_n(t)| \leq L \int_{t_0}^{t} |y(s) - x_{n-1}(s)|\, ds, \qquad t \in [t_0, t_0 + \delta]. \tag{6.10}$$

But $|y(t) - f(t)| \leq M(t - t_0)$, and using (6.10) we get

$$|y(t) - x_n(t)| \leq ML^n \frac{(t-t_0)^n}{n!}, \qquad t \in [t_0, t_0 + \delta]. \tag{6.11}$$

Hence,

$$\lim_{n\to\infty} x_n(t) = y(t) \equiv x(t), \qquad t \in [t_0, t_0 + \delta],$$

which ends the proof of Theorem 6.1.

Remark 1. Assume $f(t) = x_0 = \text{const}$, $k(t, s, x) \equiv k(s, x)$. Theorem 6.1 reduces to Theorem 2.1 for ordinary differential equations.

Remark 2. If $k(t, s, x)$ is defined on

$$\Delta': \qquad t_0 \le s \le t \le t_0 + a, \qquad |x| < \infty$$

instead of Δ, there is no need to restrict the interval $[t_0, t_0 + a]$. All $x_n(t)$ are continuous on $[t_0, t_0 + a]$, and the solution will be defined on the same interval. For instance, the linear case corresponding to $k(t, s, x) = k(t, s)x$, where $k(t, s)$ is continuous, illustrates this situation.

Remark 3. A similar theorem is valid for a vector integral equation of Volterra type. Such an equation can be written

$$x(t) = f(t) + \int_{t_0}^{t} k(t, s, x(s))\, ds \tag{6.12}$$

where x, f, and k are vector functions with the same number of components.

§6.2. An Existence Theorem

We shall consider again Eq. (6.1) and prove an existence theorem that generalizes Peano's existence theorem for ordinary differential equations (Theorem 2.4).

Theorem 6.2

Assume that $f(t)$ is a continuous function on $[t_0, t_0 + a]$, $a > 0$, and $k(t, s, x)$ is continuous on

$$\Delta: \qquad t_0 \le s \le t \le t_0 + a, \qquad |x - f(t)| \le b, \qquad b > 0.$$

Then there exists at least one continuous solution of Eq. (6.1), *defined on* $[t_0, t_0 + \delta]$, *with δ given by* (6.2).

Proof. Consider the following sequence of continuous functions on $[t_0, t_0 + \delta]$:

$$x_1(t) = f(t) \tag{6.13}$$

and for $n > 1$

$$\left.\begin{cases} x_n(t) = f(t), & t \in \left[t_0, t_0 + \dfrac{\delta}{n}\right] \\ x_n(t) = f(t) + \displaystyle\int_{t_0}^{t - (\delta/n)} k(t, s, x_n(s))\, ds, & t \in \left[t_0 + \dfrac{\delta}{n}, t_0 + \delta\right]. \end{cases}\right\} \tag{6.14}$$

These formulas should be understood as follows: The first formula in (6.14) defines $x_n(t)$ on $[t_0, t_0 + (\delta/n)]$. From the second formula in (6.14), we get

$$x_n(t) = f(t) + \int_{t_0}^{t-(\delta/n)} k(t, s, f(s))\, ds, \qquad t \in \left[t_0 + \frac{\delta}{n}, t_0 + \frac{2\delta}{n}\right].$$

The successive step will give $x_n(t)$ on the interval $[t_0 + (2\delta/n), t_0 + (3\delta/n)]$, and so on.

Let us point out that this definition of $x_n(t)$ makes sense, because

$$|x_n(t) - f(t)| \leq M\left(t - \frac{\delta}{n} - t_0\right) < M\delta \leq b. \tag{6.15}$$

It can be seen easily that all $x_n(t)$ are continuous on $[t_0, t_0 + \delta]$. From (6.15) it follows that they are uniformly bounded on the same interval. Finally, the sequence is equicontinuous, which follows from the inequality

$$|x_n(t) - x_n(s)| \leq |f(t) - f(s)| + M|t - s|$$

for any $n \geq 1$ and $t, s \in [t_0, t_0 + \delta]$.

Consequently, we can apply the lemma of Ascoli–Arzelà, which guarantees the existence of a uniformly convergent subsequence $\{x_{n_k}(t)\}$, $k = 1, 2, \ldots$. From the second equation in (6.14), considered for $n = n_k$, $k = 1, 2, \ldots$, we find for $k \to \infty$:

$$x(t) = f(t) + \int_{t_0}^{t} k(t, s, x(s))\, ds$$

where $x(t) = \lim x_{n_k}(t)$, as $k \to \infty$. Theorem 6.2 is completely proved.

Remark. The proof of Theorem 6.2 constitutes an alternate proof for Theorem 2.4.

§6.3. The Linear Equation

As mentioned, the Volterra linear equation of the second kind has the form

$$x(t) = f(t) + \int_{t_0}^{t} k(t, s)x(s)\, ds, \qquad t \in [t_0, t_0 + a], \tag{6.16}$$

where $f(t)$ and $k(t, s)$ are scalar continuous functions on $[t_0, t_0 + a]$ and $t_0 \leq s \leq t \leq t_0 + a$, respectively. The method of successive approximations is applicable (Theorem 6.1), and the following sequence can be constructed:

$$\left.\begin{aligned}
&x_0(t) = f(t),\\
&x_1(t) = f(t) + \int_{t_0}^{t} k(t, s)f(s)\, ds,\\
&x_2(t) = f(t) + \int_{t_0}^{t} k(t, s)f(s)\, ds + \int_{t_0}^{t} k_2(t, s)f(s)\, ds,\\
&\dots\dots\dots\dots\dots\dots\dots\dots\dots\dots\\
&x_n(t) = f(t) + \int_{t_0}^{t} k(t, s)f(s)\, ds + \cdots + \int_{t_0}^{t} k_n(t, s)f(s)\, ds,\\
&\dots\dots\dots\dots\dots\dots\dots\dots\dots\dots
\end{aligned}\right\} \tag{6.17}$$

where

$$k_n(t, s) = \int_{s}^{t} k(t, u)k_{n-1}(u, s)\, du, \qquad n \geq 2. \tag{6.18}$$

It is convenient to denote $k_1(t, s) = k(t, s)$.

From the general formula of recurrence,

$$x_n(t) = f(t) + \int_{t_0}^{t} k(t, s)x_{n-1}(s)\, ds$$

we obtain for $n = 2$,

$$x_2(t) = f(t) + \int_{t_0}^{t} k(t, s)f(s)\, ds + \int_{t_0}^{t} k(t, s)\, ds \int_{t_0}^{s} k(s, u)f(u)\, du$$

taking into account the second formula (6.17). But

$$\int_{t_0}^{t} k(t, s)\, ds \int_{t_0}^{s} k(s, u)f(u)\, du = \int_{t_0}^{t} f(u)\, du \int_{u}^{t} k(t, s)k(s, u)\, ds.$$

Taking (6.18) into account, we get for $x_2(t)$ the form mentioned previously. Similar arguments will show that $x_n(t)$ has the form given in (6.17). We can also write

$$x_n(t) = f(t) + \int_{t_0}^{t} \left(\sum_{j=1}^{n} k_j(t, s)\right) f(s)\, ds \tag{6.19}$$

for all $n \geq 1$, $t \in [t_0, t_0 + a]$.

Now, we shall prove that the series

$$\sum_{j=1}^{\infty} k_j(t, s) \tag{6.20}$$

is uniformly convergent on $t_0 \le s \le t \le t_0 + a$. Let $K > 0$ such that $|k(t, s)| \le K$, $t_0 \le s \le t \le t_0 + a$. From (6.18), for $n = 2$, we obtain

$$|k_2(t, s)| \le K^2(t - s), \qquad t_0 \le s \le t \le t_0 + a$$

and, using (6.18) repeatedly, we shall find

$$|k_n(t, s)| \le K^n \frac{(t - s)^{n-1}}{(n - 1)!}, \qquad t_0 \le s \le t \le t_0 + a.$$

These estimates prove that (6.20) is uniformly convergent on $t_0 \le s \le t \le t_0 + a$. Denoting by $\tilde{k}(t, s)$ the sum of the series (6.20), we get from (6.19), as $n \to \infty$,

$$x(t) = f(t) + \int_{t_0}^{t} \tilde{k}(t, s) f(s)\, ds. \tag{6.21}$$

This formula gives the solution of (6.16), and it shows that $\tilde{k}(t, s)$ serves to represent the solution for any continuous $f(t)$.

Usually, $k(t, s)$ is called the *kernel* of Eq. (6.16). The function $\tilde{k}(t, s)$ that we constructed is called the *resolvent kernel* for Eq. (6.16). It is obviously a continuous function on $t_0 \le s \le t \le t_0 + a$. Consequently, construction of the resolvent kernel allows the integral representation of the solution for any free term $f(t)$. Equation (6.21) makes sense for piecewise continuous $f(t)$ or even more general functions.

Finally, from

$$\tilde{k}(t, s) = k(t, s) + \sum_{j=2}^{\infty} k_j(t, s)$$

and (6.18), we deduce

$$\tilde{k}(t, s) = k(t, s) + \int_{t_0}^{t} k(t, u)\tilde{k}(u, s)\, du \tag{6.22}$$

which is the integral equation for the resolvent kernel.

§6.4. The First-Kind Linear Equation

The first-kind Volterra integral equation has the form

$$f(t) = \int_{t_0}^{t} k(t, s) x(s)\, ds \tag{6.23}$$

where $f(t)$ and $k(t, s)$ are known functions and $x(t)$ is to be determined. We want to show how Eq. (6.23) can be reduced to a Volterra integral equation of the second kind, if some suitable conditions hold.

But before doing so, we shall make some formal observations about this equation. The right-hand side of (6.23) defines an *operator*—that is, for any $x(t)$ belonging to a given class of functions there corresponds another function—namely, $\int_{t_0}^{t} k(t, s)x(s)\, ds$. We can write also

$$(Kx)(t) = \int_{t_0}^{t} k(t, s)x(s)\, ds \tag{6.24}$$

for any $x(t)$ such that (6.24) makes sense. This allows us to write (6.23) in the form

$$Kx = f. \tag{6.25}$$

Now it is clear that to solve Eq. (6.23), for any $f(t)$ from a given class of functions, we must find the inverse operator K^{-1}. We have mentioned these simple facts because Volterra himself considered the problem of solving integral equations as a special case of finding the inverse of an integral operator.

The precise conditions under which (6.23) can be reduced to an integral equation of form (6.16) are the following:

1. $f(t)$ is continuously differentiable on $[t_0, t_0 + a]$ and $f(t_0) = 0$;
2. $k(t, s)$ and $\partial k(t, s)/\partial t$ are continuous on $t_0 \le s \le t \le t_0 + a$, and, moreover, $k(t, t) \ne 0$ on $[t_0, t_0 + a]$.

Indeed, assuming the existence of a continuous solution for (6.23), we can take the derivatives of both sides of this equation. It follows that

$$f'(t) = k(t, t)x(t) + \int_{t_0}^{t} \frac{\partial k(t, s)}{\partial t}\, x(s)\, ds. \tag{6.26}$$

The reduction is now accomplished, because $k(t, t) \ne 0$ on $[t_0, t_0 + a]$. We can state the existence and uniqueness of the solution of Eq. (6.26). This solution is continuous, and, integrating (6.26) we find (6.23), taking into account that $f(t_0) = 0$. Consequently, the equivalence of (6.23) and (6.26) is established under conditions 1 and 2.

§6.5. Some Problems on the Half-Axis

Consider the nonlinear integral equation

$$x(t) = f(t) + \int_{t_0}^{t} k(t, s)g(s, x(s))\, ds, \qquad t \ge 0 \tag{6.27}$$

where $f(t)$ is continuous on $[0, \infty)$, $k(t, s)$ is continuous on $0 \le s \le t < \infty$, and $g(t, x)$ is continuous on $[0, \infty) \times (-\infty, \infty)$. We are interested in the existence of solutions on the whole half-axis $[0, \infty)$, and, moreover, in the behavior of such solutions (e.g., boundedness, exponential decay at infinity, etc.).

Let us remark, first of all, that our conditions do not ensure the existence of a continuous solution on the whole half-axis. For instance, the equation

$$x(t) = 1 + \int_0^t x^2(s)\, ds$$

which has the same continuous solutions as $\dot{x}(t) = x^2(t)$, $x(0) = 1$, offers an example of this kind. If we assume a Lipschitz condition for $g(t, x)$—namely,

$$|g(t, x) - g(t, y)| \leq L|x - y|, \qquad t \geq 0 \tag{6.28}$$

with x and y arbitrary real numbers—then the sequence of successive approximations

$$x_0(t) = f(t), \qquad t \geq 0$$

$$x_n(t) = f(t) + \int_0^t k(t, s)g(s, x_{n-1}(s))\, ds, \qquad t \geq 0, \qquad n \geq 1$$

is uniformly convergent on every finite interval $[0, T]$ to the unique solution (continuous on $t \geq 0$) of Eq. (6.27). We invite the reader to accomplish the proof of the preceding statement.

Of course, we cannot obtain more than continuity for the solution $x(t)$, if we claim only the properties listed for our equation. To get solutions with special properties, we have to restrict our assumptions concerning Eq. (6.27).

We shall prove now an existence theorem for bounded solutions.

Theorem 6.3

Assume that Eq. (6.27) *satisfies the following conditions*:

1. $f(t)$ *is continuous and bounded on* $[0, \infty)$;
2. $k(t, s)$ *is continuous on* $0 \leq s \leq t < \infty$, *and there exists* $M > 0$ *such that*

$$\int_0^t |k(t, s)|\, ds \leq M, \qquad t \geq 0; \tag{6.29}$$

3. $g(t, x)$ *is continuous on* $[0, \infty) \times (-\infty, \infty)$ *and satisfies the Lipschitz condition* (6.28), *where*

$$L < M^{-1}; \tag{6.30}$$

4. $g(t, 0)$ *is bounded on* $[0, \infty)$.

Then there exists a unique solution $x(t)$ *of* (6.27), *continuous and bounded on* $[0, \infty)$.

Proof. The method of successive approximations can be applied, as previously described in this section.

It is obvious that all $x_n(t)$ are continuous on $[0, \infty)$. It is easy to see that they are bounded on the half-axis. Indeed, $x_0(t) = f(t)$ is bounded by hypothesis. Assume that $x_{n-1}(t)$ is bounded. From

$$x_n(t) = f(t) + \int_0^t k(t, s)g(s, x_{n-1}(s))\, ds, \qquad t \geq 0 \tag{6.31}$$

it follows that

$$|x_n(t)| \leq |f(t)| + M \sup_{t \geq 0} |g(t, x_{n-1}(t))|, \qquad t \geq 0.$$

Taking into account that

$$|g(t, x_{n-1}(t))| \leq L\,|x_{n-1}(t)| + |g(t, 0)|, \qquad t \geq 0$$

we obtain that $x_n(t)$ is bounded. Thus, $\{x_n(t)\}$ is a sequence of bounded functions on the whole half-axis $t \geq 0$.

Relation (6.31) and its analogue corresponding to $n + 1$ will give for $n \geq 1$:

$$|x_{n+1}(t) - x_n(t)| \leq LM \sup_{t \geq 0} |x_n(t) - x_{n-1}(t)|, \qquad t \geq 0$$

or, which is the same,

$$\sup_{t \geq 0} |x_{n+1}(t) - x_n(t)| \leq LM \sup_{t \geq 0} |x_n(t) - x_{n-1}(t)|\,. \tag{6.32}$$

For $n = 1$, it can be easily seen from (6.31) that

$$\sup_{t \geq 0} |x_1(t) - x_0(t)| \leq MN \tag{6.33}$$

where

$$N = \sup_{t \geq 0} |g(t, f(t))|\,. \tag{6.34}$$

The number N is finite because

$$|g(t, f(t))| \leq L\,|f(t)| + |g(t, 0)|, \qquad t \geq 0.$$

From (6.32) and (6.33), we derive

$$\sup_{t \geq 0} |x_{n+1}(t) - x_n(t)| \leq L^n M^{n+1} N \tag{6.35}$$

which shows that the series

$$\sum_{n=1}^{\infty} [x_n(t) - x_{n-1}(t)]$$

is dominated, uniformly on $[0, \infty)$, by the series $MN \sum_0^\infty (LM)^n$. But

(6.30) guarantees the convergence of the last series, implying the uniform convergence of the sequence $\{x_n(t)\}$.

If we denote $x(t) = \lim_{n\to\infty} x_n(t)$, then $x(t)$ satisfies (6.27). It is obviously continuous on $[0, \infty)$ and bounded (for the last property, see the dominating series). The uniqueness follows in the standard way; we shall omit the details. Theorem 6.3 is thus proved.

Remark 1. A similar theorem is valid in the vector case. Namely, we can assume that x, f, g are n-vectors and k is an n by n matrix. The conditions (6.28) and (6.29) will involve the norms of the corresponding vectors or matrices.

Remark 2. Under more restrictive conditions concerning Eq. (6.27), we can prove that the solution is exponentially decreasing at infinity.

Assume, for instance, that

$$|f(t)| \leq A \exp\{-\alpha t\}, \qquad t \geq 0, \qquad \alpha > 0$$
$$|k(t, s)| \leq k_0 \exp\{-\alpha(t - s)\}, \qquad 0 \leq s \leq t < \infty$$
$$g(t, 0) = 0, \qquad t \geq 0$$

maintaining the continuity conditions and the Lipschitz condition for $g(t, x)$. Obviously

$$\int_0^t |k(t, s)|\, ds \leq k_0 \alpha^{-1} = M, \qquad t \geq 0.$$

That is, condition (6.29) is satisfied.

Consequently, there exists a unique bounded solution of Eq. (6.27). Let us prove that it is exponentially decreasing. From Eq. (6.27) we get, by virtue of the conditions listed,

$$|x(t)| \leq A \exp\{-\alpha t\} + k_0 \int_0^t \exp\{-\alpha(t - s)\} \cdot L|x(s)|\, ds$$

or

$$e^{\alpha t}|x(t)| \leq A + k_0 L \int_0^t e^{\alpha s}|x(s)|\, ds, \qquad t \geq 0. \tag{6.36}$$

This inequality is of the Gronwall type and leads easily to

$$e^{\alpha t}|x(t)| \leq A \exp(k_0 L t), \qquad t \geq 0$$

from which

$$|x(t)| \leq A \exp[-(\alpha - k_0 L)t], \qquad t \geq 0. \tag{6.37}$$

We notice that condition (6.30) now becomes $\alpha - k_0 L > 0$, which implies that $x(t)$ tends exponentially toward zero as $t \to \infty$.

In concluding this section, we would like to point out that the problems we discussed are closely related to the stability properties of the differential systems

$$\dot{x} = A(t)x + f(t, x) \tag{6.38}$$

which we studied in §5.3. As shown in §5.3, system (6.38), with the initial condition $x(t_0) = x^0$, is equivalent to the (vector) integral equation

$$x(t) = X(t)x^0 + \int_0^t X(t)X^{-1}(s)f(s, x(s))\, ds \tag{6.39}$$

where $\dot{X} = A(t)X$, $X(0) = I$. Equation (6.39) has the form (6.27), and it is now obvious that the preceding result concerning the existence of solutions with exponential decay at infinity is intimately related to the result expressed by Theorem 5.4.

PROBLEMS

1. Consider the Volterra integral equation with the Stieltjes integral

$$x(t) = f(t) + \int_{t_0}^t k(t, s, x(s))\, d\sigma(s).$$

Find conditions that assure the existence and uniqueness of the solution. Compare with the case $\sigma(t) \equiv t$.

2. Let $u(t)$ and $f(t)$ be scalar continuous functions, nonnegative on $t_0 \le t \le t_0 + T$. Assume further that $k(t, s)$ is continuous and nonnegative on $t_0 \le s \le t \le t_0 + T$. Then

$$u(t) \le f(t) + \int_{t_0}^t k(t, s)u(s)\, ds, \qquad t_0 \le t \le t_0 + T$$

implies

$$u(t) \le f(t) + \int_{t_0}^t \tilde{k}(t, s)f(s)\, ds, \qquad t_0 \le t \le t_0 + T$$

where $\tilde{k}(t, s)$ is the resolvent kernel associated with $k(t, s)$.

3. Use the preceding result to obtain Gronwall's inequality (see §1.5).

4. Consider the Volterra integral equation $x(t) = f(t) + \int_0^t k(t-s)x(s)\, ds$. Prove that the resolvent kernel depends only on the difference $t - s$.

5. Find a Volterra integral equation that is equivalent to the differential equation

$$x^{(n)} + a_1(t)x^{(n-1)} + \cdots + a_n(t)x = f(t)$$

with the usual initial conditions (at t_0).

Hint: Denote the nth derivative of $x(t)$ by $y(t)$. Then

$$x^{(n-1)}(t) = x_0^{(n-1)} + \int_0^t (t-s)y(s)\,ds,$$

$$x^{(n-2)}(t) = x_0^{(n-2)} + tx_0^{(n-1)} + \int_0^t \frac{(t-s)^2}{2!}\,y(s)\,ds, \ldots,$$

$$x(t) = x_0 + tx_0' + \cdots + \frac{t^{n-1}}{(n-1)!}\,x_0^{(n-1)} + \int_0^t \frac{(t-s)^{n-1}}{(n-1)!}\,y(s)\,ds.$$

Now, replace $x(t)$ and its derivatives in the differential equation.

6. Discuss the Volterra equation in two independent variables

$$x(t, s) = f(t, s) + \int_0^t \int_0^s k(t, s; u, v)x(u, v)\,du\,dv.$$

7

Fredholm Theory of Linear Integral Equations

One of the most beautiful theories of mathematical analysis, with important consequences in the development of modern mathematics, is undoubtedly that formulated by the Swedish mathematician Ivar Fredholm at the beginning of this century. It concerns linear integral equations with fixed range of integration and containing a complex parameter. The presence of the parameter makes necessary a discussion of the existence of solutions, which represents the elegant feature of the theory.

We are not going to present this theory in its full extent. The restriction to the case of equations with continuous kernel allows us to avoid the use of Lebesgue integral, which is rather difficult for some nonmathematician readers. On the other hand, we would like to point out that this restriction is not essential, because in the Fredholm theory of equations with continuous kernel we shall find all the significant features of the general theory (as it can be found, for instance, in many books on functional analysis).

§7.1. The Resolvent Kernel

Consider the linear integral equation of Fredholm type

$$x(t) = f(t) + \lambda \int_a^b k(t, s)x(s)\, ds, \qquad t \in [a, b] \tag{7.1}$$

where $f(t)$ is a continuous complex-valued function on $[a, b]$, $k(t, s)$ is a continuous complex-valued function on $[a, b] \times [a, b]$, and λ is a complex parameter. A first result concerning the existence of solutions of Eq. (7.1) can be obtained by the successive approximations method.

Theorem 7.1

Equation (7.1) *has a unique continuous solution on* $[a, b]$ *if*

$$|\lambda| M(b-a) < 1 \tag{7.2}$$

where

$$M = \sup |k(t, s)|, \qquad t, s \in [a, b]. \tag{7.3}$$

Proof. Let us define the sequence of continuous functions on $[a, b]$

$$x_0(t) = f(t),$$

$$x_1(t) = f(t) + \lambda \int_a^b k(t, s) f(s)\, ds,$$

$$\cdots\cdots\cdots\cdots\cdots\cdots\cdots\cdots$$

$$x_n(t) = f(t) + \lambda \int_a^b k(t, s) x_{n-1}(s)\, ds,$$

$$\cdots\cdots\cdots\cdots\cdots\cdots\cdots\cdots$$

It can be easily seen that

$$x_n(t) = f(t) + \lambda \int_a^b k(t, s) f(s)\, ds + \cdots + \lambda^n \int_a^b k_n(t, s) f(s)\, ds \tag{7.4}$$

where

$$k_n(t, s) = \int_a^b k(t, u) k_{n-1}(u, s)\, du, \qquad n \geq 2. \tag{7.5}$$

From (7.5) it follows that

$$k_n(t, s) = \underbrace{\int_a^b \cdots \int_a^b}_{(n-1)} k(t, t_1) k(t_1, t_2) \cdots k(t_{n-1}, s)\, dt_1 \cdots dt_{n-1} \tag{7.6}$$

which leads to the following formula:

$$k_n(t, s) = \int_a^b k_p(t, u) k_{n-p}(u, s)\, du, \qquad 1 \leq p \leq n-1 \tag{7.7}$$

where $k_1(t, s) = k(t, s)$ by definition.

Formula (7.4) can be also written

$$x_n(t) = f(t) + \lambda \int_a^b \left\{ \sum_{j=1}^{n} k_j(t, s)\lambda^{j-1} \right\} f(s)\, ds, \qquad n \geq 1. \tag{7.8}$$

But the series

$$\sum_{j=1}^{\infty} k_j(t, s)\lambda^{j-1} \tag{7.9}$$

is uniformly convergent with respect to $(t, s) \in [a, b] \times [a, b]$, for λ satisfying (7.2).

Indeed, from (7.6) we obtain

$$|k_n(t, s)| \leq M^n(b-a)^{n-1}, \qquad n \geq 1 \tag{7.10}$$

which shows that the series (7.9) is dominated by the series having the general term $M^n(|\lambda|(b-a))^{n-1}$—that is, by a geometric convergent series. If we denote

$$R(t, s; \lambda) = \sum_{j=1}^{\infty} k_j(t, s)\lambda^{j-1} \tag{7.11}$$

it follows that $R(t, s; \lambda)$ is a continuous function in $(t, s; \lambda)$, for $(t, s) \in [a, b] \times [a, b]$ and $|\lambda| < [M(b-a)]^{-1}$.

From (7.4) and its analogue for $n-1$, one obtains

$$|x_n(t) - x_{n-1}(t)| \leq N(|\lambda| M(b-a))^n, \qquad n \geq 1 \tag{7.12}$$

which shows that the sequence $\{x_n(t)\}$ is uniformly convergent on $[a, b]$ (we put $N = \sup |f(t)|$). Taking (7.8) and (7.11) into account, we get

$$x(t) = f(t) + \lambda \int_a^b R(t, s; \lambda) f(s)\, ds \tag{7.13}$$

where $x(t)$ denotes the limit of the sequence $\{x_n(t)\}$. From the construction of this sequence, it follows that $x(t)$ is a solution of Eq. (7.1). Consequently, (7.13) gives a solution of Eq. (7.1), under condition (7.2).

To prove the uniqueness, assume that $y(t)$ is a solution of the same equation:

$$y(t) = f(t) + \lambda \int_a^b k(t, s) y(s)\, ds.$$

Using the recurrence formula for $x_n(t)$, we obtain

$$|y(t) - x_n(t)| \leq |\lambda| \int_a^b |k(t, s)| \cdot |y(s) - x_{n-1}(s)|\, ds, \qquad n \geq 1. \tag{7.14}$$

If we denote $\alpha_n = \sup |y(t) - x_n(t)|$, $t \in [a, b]$, then (7.14) leads to

$$\alpha_n \leq |\lambda| M(b-a)\alpha_{n-1}, \qquad n \geq 1 \tag{7.15}$$

and, finally,

$$\alpha_n \le (M|\lambda|(b-a))^n \alpha_0, \qquad n \ge 1. \tag{7.16}$$

Condition (7.2) shows that $\alpha_n \to 0$ as $n \to \infty$—that is, $y(t) = \lim x_n(t) = x(t)$, $t \in [a, b]$. Theorem 7.1 is thus proved.

Remark 1. The function $R(t, s; \lambda)$ is called the *resolvent kernel* associated with $k(t, s)$. From (7.13), it follows that finding the resolvent kernel or solving Eq. (7.1), for any continuous $f(t)$, are equivalent problems.

Remark 2. Formula (7.11) shows that $R(t, s; \lambda)$ is an analytic function with respect to λ, $|\lambda| < [M(b-a)]^{-1}$.

Remark 3. The kernel $k(t, s)$ and the resolvent kernel $R(t, s; \lambda)$ are related by the following equations:

$$R(t, s; \lambda) = k(t, s) + \lambda \int_a^b k(t, u)R(u, s; \lambda)\, du \tag{7.17}$$

$$R(t, s; \lambda) = k(t, s) + \lambda \int_a^b R(t, u; \lambda)k(u, s)\, du. \tag{7.18}$$

Indeed, from (7.11) and (7.7) we can obtain both relations (7.17) and (7.18) by very simple transformations. For instance, (7.11) can be written in the form

$$R(t, s; \lambda) = k(t, s) + \lambda \int_a^b k(t, u)\{k(u, s) + \lambda k_2(u, s) + \cdots\}\, du$$

which is the same as (7.17). Each equation (7.17) or (7.18) defines completely the resolvent kernel for those values of λ given by (7.2). The significant role of Eqs. (7.17) and (7.18) will follow from the next theorem.

Theorem 7.2

Assume that there exists a function $R(t, s; \lambda)$, *for a given* λ [*not necessarily satisfying* (7.2)], *continuous for* $(t, s) \in [a, b] \times [a, b]$ *and such that* (7.17) *and* (7.18) *are verified. Then* (7.1) *has a unique solution for this value of* λ, *and it is given by* (7.13).

Proof. First, let us prove that $x(t)$, given by (7.13), satisfies Eq. (7.1). Indeed, if we replace $x(t)$ in (7.1) by the expression (7.13), we find

$$f(t) + \lambda \int_a^b R(t, s; \lambda)f(s)\, ds$$

$$= f(t) + \lambda \int_a^b k(t, s)\left[f(s) + \lambda \int_a^b R(s, u; \lambda)f(u)\, du\right] ds.$$

This can be written also

$$\int_a^b R(t, s; \lambda) f(s)\, ds - \int_a^b k(t, s) f(s)\, ds$$
$$- \lambda \int_a^b \int_a^b k(t, s) R(s, u; \lambda) f(u)\, ds\, du = 0$$

and finally

$$\int_a^b \left\{ R(t, s; \lambda) - k(t, s) - \lambda \int_a^b k(t, u) R(u, s; \lambda)\, du \right\} f(s)\, ds = 0$$

which is obviously true [see (7.17)].

To prove the uniqueness of the solution, we shall derive the formula (7.13) from (7.1) and (7.18). Indeed, assume that $y(t)$ is a continuous solution of (7.1):

$$y(s) = f(s) + \lambda \int_a^b k(s, u) y(u)\, ds.$$

Now, let us multiply this equation by $\lambda R(t, s; \lambda)$ and integrate both sides with respect to s. We obtain

$$\lambda \int_a^b R(t, s; \lambda) y(s)\, ds = \lambda \int_a^b R(t, s; \lambda) f(s)\, ds$$
$$+ \lambda \int_a^b y(u)\, du \left[\lambda \int_a^b R(t, s; \lambda) k(s, u)\, ds \right].$$

But (7.18) gives

$$\lambda \int_a^b R(t, s; \lambda) k(s, u)\, ds = R(t, u; \lambda) - k(t, u)$$

which shows that the preceding formula can be written

$$\lambda \int_a^b R(t, s; \lambda) y(s)\, ds = \lambda \int_a^b R(t, s; \lambda) f(s)\, ds$$
$$+ \lambda \int_a^b R(t, u; \lambda) y(u)\, du - \lambda \int_a^b k(t, u) y(u)\, du.$$

Taking into account that

$$\lambda \int_a^b k(t, u) y(u)\, du = y(t) - f(t)$$

[see (7.1)], we obtain for $y(t)$ the same expression as for $x(t)$—that is, the formula (7.13). Theorem 7.2 is thus completely proved. It shows the importance of constructing a function $R(t, s; \lambda)$, with the required properties.

§7.2. The Entire Functions of Fredholm and Their Applications

To accomplish the construction of a function with the properties of the resolvent kernel, we shall define the entire functions $D(\lambda)$ and $D(t, s; \lambda)$. For a given kernel $k(t, s)$, continuous on $[a, b] \times [a, b]$, let us consider the function

$$k\begin{pmatrix} t_1 t_2 \cdots t_n \\ s_1 s_2 \cdots s_n \end{pmatrix} = \begin{vmatrix} k(t_1, s_1) & k(t_1, s_2) \cdots k(t_1, s_n) \\ k(t_2, s_1) & k(t_2, s_2) \cdots k(t_2, s_n) \\ & \vdots \\ k(t_n, s_1) & k(t_n, s_2) \cdots k(t_n, s_n) \end{vmatrix}.$$

We can now define the following quantities:

$$d_n = \underbrace{\int_a^b \cdots \int_a^b}_{n} k\begin{pmatrix} t_1 t_2 \cdots t_n \\ t_1 t_2 \cdots t_n \end{pmatrix} dt_1\, dt_2 \cdots dt_n \tag{7.19}$$

$$d_n(t, s) = \underbrace{\int_a^b \cdots \int_a^b}_{n} k\begin{pmatrix} t\ t_1 \cdots t_n \\ s\ t_1 \cdots t_n \end{pmatrix} dt_1\, dt_2 \cdots dt_n. \tag{7.20}$$

Consider the entire series in λ

$$D(\lambda) = 1 + \sum_{n=1}^{\infty} (-1)^n \frac{\lambda^n}{n!} d_n \tag{7.21}$$

$$D(t, s; \lambda) = k(t, s) + \sum_{n=1}^{\infty} (-1)^n \frac{\lambda^n}{n!} d_n(t, s) \tag{7.22}$$

where d_n and $d_n(t, s)$ are given by (7.19) and (7.20). To prove that the series in (7.21) and (7.22) are convergent for any complex λ, we shall use the Hadamard's estimate for the modulus of a determinant. Namely, if $A = (a_{ij})$ is a square matrix of order n, such that $|a_{ij}| \leq M$, $i, j = 1, 2, \ldots, n$, then $|\det A| \leq n^{n/2} M^n$.

We find the following estimates for d_n and $d_n(t, s)$:

$$|d_n| \leq n^{n/2}[M(b-a)]^n \tag{7.23}$$

$$|d_n(t, s)| \leq (n+1)^{(n+1)/2} M^{n+1} (b-a)^n. \tag{7.24}$$

Consequently, the series in (7.21) is dominated by the series

$$1 + \sum_{n=1}^{\infty} \frac{(|\lambda|\, M(b-a))^n}{n!} n^{n/2}$$

which is convergent for any complex λ. Indeed, the ratio test can be applied.

A similar argument holds for the series in (7.22), if we take (7.24) into account. Moreover, the convergence is uniform with respect to $(t, s) \in [a, b] \times [a, b]$. This means that $D(t, s; \lambda)$ is a continuous function with respect to $(t, s) \in [a, b] \times [a, b]$.

We shall now establish a relation involving $D(\lambda)$, $D(t, s; \lambda)$, and $R(t, s; \lambda)$, for λ satisfying (7.2). This relation will serve to extend $R(t, s; \lambda)$ to the whole complex plane (as a function of λ).

Theorem 7.3

If λ satisfies (7.2), *then*

$$R(t, s; \lambda) D(\lambda) = D(t, s; \lambda) \tag{7.25}$$

for $(t, s) \in [a, b] \times [a, b]$.

Proof. Consider Eq. (7.17) and multiply both sides by $D(\lambda)$. We obtain

$$R(t, s; \lambda)D(\lambda) = k(t, s)D(\lambda) + \lambda \int_a^b k(t, u)R(u, s; \lambda)D(\lambda)\, du. \tag{7.26}$$

To establish (7.25), it suffices to prove that

$$D(t, s; \lambda) = k(t, s)D(\lambda) + \lambda \int_a^b k(t, u)D(u, s; \lambda)\, du \tag{7.27}$$

for any λ satisfying (7.2).

We shall prove (7.27) by direct calculation. If we take (7.21) and (7.22) into account, (7.27) becomes

$$k(t, s) + \sum_{n=1}^{\infty} (-1)^n \frac{\lambda^n}{n!} d_n(t, s) = k(t, s) + \sum_{n=1}^{\infty} (-1)^n \frac{\lambda^n}{n!} k(t, s)\, d_n$$
$$+ \lambda \int_a^b k(t, u)k(u, s)\, du + \sum_{n=1}^{\infty} (-1)^n \frac{\lambda^{n+1}}{n!} \int_a^b k(t, u)\, d_n(u, s)\, du.$$

This is obviously satisfied if

$$d_n(t, s) = k(t, s)\, d_n - n \int_a^b k(t, u)\, d_{n-1}(u, s)\, du \tag{7.28}$$

for $n \geq 1$, where $d_0(t, s) = k(t, s)$ by definition. Let us remark that

$$k\begin{pmatrix} t\ t_1 \cdots t_n \\ s\ t_1 \cdots t_n \end{pmatrix} = k(t, s)k\begin{pmatrix} t_1 \cdots t_n \\ t_1 \cdots t_n \end{pmatrix}$$
$$- k(t, t_1)k\begin{pmatrix} t_1 t_2 \cdots t_n \\ s\ t_2 \cdots t_n \end{pmatrix} - \cdots - k(t, t_n)k\begin{pmatrix} t_1 \cdots t_n \\ t_1 \cdots s \end{pmatrix}$$

which can be obtained easily from the definition of

$$k\begin{pmatrix} t_1 & \cdots & t_n \\ s_1 & \cdots & s_n \end{pmatrix}.$$

Integrating both sides from a to b, with respect to $t_1, \ldots, t_n$, we obtain

$$\begin{aligned} d_n(t, s) &= k(t, s)\, d_n - \sum_{i=1}^{n} \int_a^b k(t, t_i)\, d_{n-1}(t_i, s)\, dt_i \\ &= k(t, s)\, d_n - n \int_a^b k(t, u)\, d_{n-1}(u, s)\, du \end{aligned}$$

that is, Eq. (7.28). Consequently, Theorem 7.3 is proved.

The particular significance of this theorem consists in the fact that it allows us to define the function $R(t, s; \lambda)$ for λ not necessarily satisfying (7.2). Namely, let us define $R(t, s; \lambda)$ by

$$R(t, s; \lambda) = \frac{D(t, s; \lambda)}{D(\lambda)} \tag{7.29}$$

for those complex λ with $D(\lambda) \neq 0$. Obviously, $R(t, s; \lambda)$ is continuous with respect to $(t, s) \in [a, b] \times [a, b]$, for all λ's considered previously. Moreover, $R(t, s; \lambda)$ satisfies both Eqs. (7.17) and (7.18). For instance, (7.17) follows from (7.27) if we divide the latter by $D(\lambda)$. To get (7.18), we can start from

$$D(t, s; \lambda) = k(t, s)D(\lambda) + \lambda \int_a^b D(t, u; \lambda)k(u, s)\, du$$

which can be established in the same manner as (7.27).

Theorem 7.4

If λ is such that $D(\lambda) \neq 0$, then Eq. (7.1) *has a unique continuous solution for any continuous $f(t)$. This solution is given by* (7.13), *where $R(t, s; \lambda)$ is defined by* (7.29).

The proof of this theorem is a direct consequence of the previous considerations and of Theorem 7.2. In order to be able to study Eq. (7.1) for those values of λ with $D(\lambda) = 0$, we need the following lemma.

Lemma

Assume that λ_0 is a zero of $D(\lambda)$ of order $k > 0$. Then, the order of λ_0 as a zero of $D(t, s; \lambda)$ is less than k.

Proof. Let $D(\lambda) = (\lambda - \lambda_0)^k D_1(\lambda)$ be the standard factorization of $D(\lambda)$, with $D_1(\lambda_0) \neq 0$.

Assume $D(t, s; \lambda) = (\lambda - \lambda_0)^l D_1(t, s; \lambda)$ with $D_1(t, s; \lambda_0) \not\equiv 0$. We shall prove that $l < k$. This statement will be the consequence of the following relation:

$$D'(\lambda) = -\int_a^b D(s, s; \lambda)\, ds. \tag{7.30}$$

We can obtain this equation easily from the definition of the entire functions $D(\lambda)$ and $D(t, s; \lambda)$, taking into account that $d_{n+1} = \int_a^b d_n(s, s)\, ds$.

Now, the left-hand side of (7.30) has λ_0 as a zero of order $k - 1$. The right-hand side has λ_0 as a zero of order at least l (because, after integration, another factor $\lambda - \lambda_0$ could appear). Hence, $l \leq k - 1$, which proves the lemma.

Remark. The lemma shows that $R(t, s; \lambda)$ defined by (7.29), considered on the whole complex plane, is a meromorphic function.

We shall prove now a result concerning the homogeneous equation associated with (7.1):

$$x(t) = \lambda \int_a^b k(t, s)x(s)\, ds. \tag{7.31}$$

First, let us remark that for any λ such that $D(\lambda) \neq 0$, Eq. (7.31) has only the zero solution $x(t) \equiv 0$. This statement follows from Theorem 7.4, because (7.31) is Eq. (7.1) for $f(t) \equiv 0$.

For those values of λ that are zeros of the function $D(\lambda)$, the following result holds.

Theorem 7.5

If λ is a zero of $D(\lambda)$, then Eq. (7.31) *has nonzero solutions.*

Proof. Assume that λ_0 is a zero of $D(\lambda)$. We know that λ_0 is a pole for $R(t, s; \lambda)$. Consequently, the following representation holds for $|\lambda - \lambda_0|$ sufficiently small:

$$R(t, s; \lambda) = \sum_{j=0}^{\infty} a_j(t, s)(\lambda - \lambda_0)^j + \sum_{i=1}^{r} \frac{b_i(t, s)}{(\lambda - \lambda_0)^i}$$

where $a_j(t, s)$ and $b_i(t, s)$ are continuous functions in the square $[a, b] \times [a, b]$ and, moreover, $b_r(t, s) \not\equiv 0$.

Consider now Eq. (7.17) for $R(t, s; \lambda)$, and let us replace in this equation $R(t, s; \lambda)$ by the preceding expansion. We get

$$\sum_{j=0}^{\infty} a_j(t, s)(\lambda - \lambda_0)^j + \sum_{i=1}^{r} \frac{b_i(t, s)}{(\lambda - \lambda_0)^i}$$

$$= k(t, s) + \lambda \int_a^b k(t, u) \left\{ \sum_{j=0}^{\infty} a_j(u, s)(\lambda - \lambda_0)^j + \sum_{i=1}^{r} \frac{b_i(u, s)}{(\lambda - \lambda_0)^i} \right\} du.$$

Multiplying both sides by $(\lambda - \lambda_0)^r$ and making $\lambda \to \lambda_0$, we obtain

$$b_r(t, s) = \lambda_0 \int_a^b k(t, u) b_r(u, s)\, du. \tag{7.32}$$

Inasmuch as $b_r(t, s) \not\equiv 0$, we can find at least one s such that $b_r(t, s) \not\equiv 0$, as a function of t, $t \in [a, b]$. This means that Eq. (7.31) has nonzero solutions for $\lambda = \lambda_0$, which proves Theorem 7.5

Before proceeding further in discussing Eq. (7.1), we formulate the following questions:

1. If λ is a zero of $D(\lambda)$, how many linear independent solutions has Eq. (7.31)?
2. What can we say about Eq. (7.1) for such values of λ?

To answer these questions, we need some knowledge related to the concept of Hilbert space. We do not intend to give an abstract presentation. Our basic set will be that of continuous complex-valued functions on $[a, b]$. It is commonly denoted by $C[a, b]$.

§7.3. A Glimpse of Hilbert Space Theory

We can easily see that $C[a, b]$ is a vector space over the field of complex numbers—that is, it is an abelian group with respect to the usual operation of addition—and a scalar multiplication is defined with complex numbers, such that:

$$\alpha[x(t) + y(t)] = \alpha x(t) + \alpha y(t)$$
$$(\alpha + \beta)x(t) = \alpha x(t) + \beta x(t)$$
$$(\alpha\beta)x(t) = \alpha[\beta x(t)]$$
$$1 \cdot x(t) = x(t).$$

These relations hold for any complex α, β and for any $x(t), y(t) \in C[a, b]$.

These two operations completely define the algebraic structure of $C[a, b]$. To be able to consider some topological concepts related with $C[a, b]$, primarily that of convergence, we shall define the *scalar product* or *inner product* on $C[a, b]$. Namely, for any $x(t), y(t) \in C[a, b]$, we define

$$(x, y) = \int_a^b x(t)\overline{y(t)}\, dt. \tag{7.33}$$

The complex number (x, y) is the *scalar product* of $x(t)$ and $y(t)$. To speak more precisely, the scalar product is a complex-valued function on $C[a, b] \times C[a, b]$, such that the following basic properties hold:

$$\begin{aligned} (x, y) &= \overline{(y, x)} \\ (x + y, z) &= (x, z) + (y, z) \\ (\alpha x, y) &= \alpha(x, y) \\ (x, x) &\geq 0 \end{aligned} \tag{7.34}$$

the equal sign implying $x = 0$.

In the case of abstract Hilbert spaces, properties (7.34) are taken as axioms for the scalar product. The scalar product allows us to define the *norm* of an arbitrary element from $C[a, b]$. The norm of $x(t) \in C[a, b]$ is given by

$$\|x\| = (x, x)^{1/2} = \left\{\int_a^b |x(t)|^2\, dt\right\}^{1/2}. \tag{7.35}$$

It is obvious that $\|x\| \geq 0$ for any $x \in C[a, b]$, the equal sign being possible only for $x = 0$ (i.e., the zero function). It is also obvious that $\|\alpha x\| = |\alpha|\, \|x\|$, for any complex α and $x \in C[a, b]$. Another important property of the norm is expressed by the so-called *triangle inequality*:

$$\|x + y\| \leq \|x\| + \|y\| \tag{7.36}$$

which holds for any $x, y \in C[a, b]$.

To prove (7.36), we shall establish an inequality involving both scalar product and norm:

$$|(x, y)| \leq \|x\|\, \|y\| \tag{7.37}$$

for any $x, y \in C[a, b]$. Inequality (7.37) is called the *Schwarz inequality*. We have

$$\|x + \lambda y\|^2 \geq 0 \tag{7.38}$$

for any complex λ and any $x, y \in C[a, b]$. We can also write (7.38) in the form

$$\|x\|^2 + \lambda(y, x) + \bar{\lambda}(x, y) + |\lambda|^2 \|y\|^2 \geq 0. \tag{7.39}$$

For $y = 0$, inequality (7.37) is obviously valid.

Assume now that $y \neq 0$ and take

$$\lambda - -\frac{(x, y)}{\|y\|^2}.$$

Equation (7.39) now becomes

$$\|x\|^2 - 2\frac{|(x, y)|^2}{\|y\|^2} + \frac{|(x, y)|^2}{\|y\|^2} \geq 0$$

which yields $\|x\|^2\|y\|^2 - |(x, y)|^2 \geq 0$—that is, (7.37). Consider now

$$\|x + y\|^2 = (x + y, x + y) = \|x\|^2 + (x, y) + (y, x) + \|y\|^2$$
$$= \|x\|^2 + 2\,\mathrm{Re}(x, y) + \|y\|^2 \leq \|x\|^2 + 2\|x\|\,\|y\| + \|y\|^2 = (\|x\| + \|y\|)^2$$

because $\mathrm{Re}(x, y) \leq |(x, y)| \leq \|x\| \cdot \|y\|$. From the preceding inequalities, it follows (7.36).

The norm of an element of $C[a, b]$ can be interpreted geometrically as the distance from 0 to that element. Accordingly, $\|x - y\|$ can be considered as the distance from x to y: $d(x, y) = \|x - y\|$. The usual properties of the distance (metric) are verified by d, which allows us to derive for $C[a, b]$ a metric structure, beside that of inner-product space.

According to the general theory of metric spaces, we shall say that the sequence $\{x_n\} \subset C[a, b]$ is convergent in the norm to $x \in C[a, b]$, if $\|x_n - x\| \to 0$ as $n \to \infty$—that is,

$$\lim_{n \to \infty} \int_a^b |x_n(t) - x(t)|^2\, dt = 0. \tag{7.40}$$

One says also that $\{x_n(t)\}$ converges *in the mean* to $x(t)$. This kind of convergence is different from that of uniform convergence.

Indeed, if $x_n(t) \to x(t)$ as $n \to \infty$, uniformly on $[a, b]$, then for any $\varepsilon > 0$ we can find $N(\varepsilon) > 0$, such that

$$|x_n(t) - x(t)| < \frac{\varepsilon}{(b - a)^{1/2}}, \qquad t \in [a, b], \qquad n \geq N(\varepsilon).$$

This implies $\|x_n - x\| < \varepsilon$ for $n \geq N(\varepsilon)$, which means that $x_n \to x$ in the mean. Consequently, any uniformly convergent sequence is convergent in the mean.

The converse of this statement is not necessarily true, as follows from the example we shall now construct:

$$x_n(t) = \begin{cases} nt & 0 \leq t \leq n^{-1} \\ 2 - nt & n^{-1} < t \leq 2n^{-1} \\ 0 & 2n^{-1} < t \leq 1 \end{cases}$$

with $a = 0$, $b = 1$. We see that $\|x_n\| \to 0$ as $n \to \infty$—that is, $\{x_n\}$ converges in the mean to $x = 0$. On the other hand, $\lim_{n \to \infty} x_n(t) = 0$ for any $t \in [0, 1]$. But $x_n(n^{-1}) = 1$ for any n, which shows that $\{x_n\}$ is not uniformly convergent to zero.

The uniqueness of the limit, with respect to the convergence in the mean, follows from the fact that in any metric space the limit of a convergent sequence is unique. We can prove this statement directly if we take into account that $\|x - y\| \leq \|x - x_n\| + \|x_n - y\|$, and $\|x_n - x\| \to 0$,

$\|x_n - y\| \to 0$ as $n \to \infty$, would imply $\|x - y\| = 0$—that is, $x = y$. Furthermore, $C[a, b]$ is not *complete* with respect to the norm—that is, from $\|x_n - x_m\| \to 0$ as $n, m \to \infty$, does not necessarily follow the existence of an element $x \in C[a, b]$, such that $\|x_n - x\| \to 0$ as $n \to \infty$. We would like to point out that the completeness is required in the definition of a Hilbert space. An inner-product space that is incomplete is also called a *pre-Hilbert space.*

The concept of scalar product allows us to define the *orthogonality* in $C[a, b]$. We shall say that x and y are orthogonal if $(x, y) = 0$. This definition is justified by the fact that $(x, y) = 0$ implies $\|x + y\|^2 = \|x\|^2 + \|y\|^2$ —that is, the Pythagorean theorem holds.

Consider now a sequence $\{x_n\} \in C[a, b]$, such that $(x_n, x_m) = 0$ for $n \neq m$ and $(x_n, x_n) = 1$. Such a sequence is called *orthonormal.* Our aim is to prove the *Bessel inequality: for any* $x \in C[a, b]$,

$$\sum_{n=1}^{\infty} |(x, x_n)|^2 \leq \|x\|^2 \tag{7.41}$$

where $\{x_n\}$ *is an orthonormal sequence.* We shall prove that

$$\sum_{n=1}^{N} |(x, x_n)|^2 \leq \|x\|^2 \tag{7.42}$$

holds for any N, which implies (7.41).

Indeed, we have with $c_n = (x, x_n)$,

$$\left\| x - \sum_{n=1}^{N} c_n x_n \right\|^2 \geq 0$$

which is equivalent to

$$\|x\|^2 - \sum_{n=1}^{N} c_n(x_n, x) - \sum_{n=1}^{N} c_n(x, x_n) + \sum_{n=1}^{N} \sum_{m=1}^{N} c_n \bar{c}_m(x_n, x_m) \geq 0.$$

But $c_n(x_n, x) = \overline{c_n(x, x_n)} = c_n\bar{c}_n = |c_n|^2$ and $\bar{c}_n(x, x_n) = |c_n|^2$. If we consider also that $\{x_n\}$ is orthonormal, the inequality becomes

$$\|x\|^2 - \sum_{n=1}^{N} |c_n|^2 \geq 0$$

that is, inequality (7.42). The quantities (x, x_n) are usually called the *Fourier coefficients* of x, with respect to the system $\{x_n\}$.

Later, we shall need the following simple property of finite-dimensional spaces of functions.

Any finite-dimensional space of continuous functions on $[a, b]$ *has an orthonormal basis.*

Indeed, let $\{\psi_k(t)\}$, $k = 1, 2, \ldots, n$ be a basis of the given space. Let us denote

$$\varphi_1(t) = \frac{\psi_1(t)}{\|\psi_1\|},$$

$$\tilde{\varphi}_2(t) = \psi_2(t) - (\psi_2, \varphi_1)\varphi_1(t), \qquad \varphi_2(t) = \frac{\tilde{\varphi}_2(t)}{\|\tilde{\varphi}_2\|}, \ldots,$$

$$\tilde{\varphi}_k(t) = \psi_k(t) - (\psi_k, \varphi_{k-1})\varphi_{k-1}(t) - \cdots - (\psi_k, \varphi_1)\varphi_1(t), \qquad \varphi_k(t) = \frac{\tilde{\varphi}_k(t)}{\|\tilde{\varphi}_k\|}$$

for $k \leq n$. Obviously, $\varphi_k(t)$, $k = 1, 2, \ldots, n$ are linearly independent (otherwise, the $\psi_k(t)$ would be linearly dependent) and $(\varphi_k, \varphi_j) = 0$ for $j \neq k$, $(\varphi_k, \varphi_k) = 1$. The preceding procedure to construct an orthonormal basis is known as *Schmidt orthogonalization method*.

We are now able to develop further the discussion of Eq. (7.1) and to answer completely the questions formulated at the end of §7.2.

§7.4. Eigenvalues, Eigenfunctions and Adjoint Equations

Theorem 7.4 completely clarifies the situation for those values of λ for which $D(\lambda) \neq 0$. In such cases, Eq. (7.1) has a unique solution for any continuous free term $f(t)$, and, in particular, the corresponding homogeneous equation (7.31) has only the zero solution $x(t) \equiv 0$.

The values of λ such that $D(\lambda) = 0$ are called eigenvalues of the kernel $k(t, s)$.

Note that zero cannot be an eigenvalue. Inasmuch as $D(\lambda)$ is an analytic entire function, it follows that *the set of eigenvalues of a continuous kernel is at most countable*. Moreover, *if the set of eigenvalues is infinite, the unique limit point of this set is ∞.*

Theorem 7.5 shows that for any eigenvalue λ, the homogeneous equation (7.31) has nonzero solutions.

Any nonzero solution of Eq. (7.31) *is called an eigenfunction of the kernel $k(t, s)$, corresponding to the eigenvalue λ.*

From these considerations, λ is obviously an eigenvalue of the kernel $k(t, s)$ if and only if the homogeneous equation (7.31) has nonzero solutions. This remark allows us to define the eigenvalues as being those values of λ for which Eq. (7.31) has nonzero solutions.

We can complete now the discussion of Eq. (7.31), when λ is an eigenvalue of the kernel $k(t, s)$.

Theorem 7.6

Assume that λ is an eigenvalue of $k(t, s)$. Then, the set of eigenfunctions corresponding to λ is a linear space of finite dimension. The dimension of this space is called the multiplicity of the eigenvalue λ.

Proof. That the set of eigenfunctions is a linear space is very easy to prove. To prove that its dimension is finite, we consider a set of n linearly independent eigenfunctions corresponding to λ. We shall obtain an upper estimate for n, which shows that the maximum number of linearly independent elements belonging to this space is finite.

Without loss of generality, we can assume that $\varphi_1(t), \ldots, \varphi_n(t)$ form an orthonormal system (applying, if necessary, the Schmidt procedure of orthogonalization). Let us write that these functions are eigenfunctions corresponding to λ:

$$\varphi_j(t) = \lambda \int_a^b k(t, s)\varphi_j(s)\, ds, \qquad j = 1, 2, \ldots, n. \tag{7.43}$$

We can write (7.43) in the equivalent form

$$\frac{1}{\bar{\lambda}} \overline{\varphi_j(t)} = \int_a^b \overline{k(t, s)}\overline{\varphi_j(s)}\, ds, \qquad j = 1, 2, \ldots, n \tag{7.44}$$

which means that

$$\frac{1}{\bar{\lambda}} \overline{\varphi_j(t)} = (\overline{k(t, s)}, \varphi_j(s)), \qquad j = 1, 2, \ldots, n. \tag{7.45}$$

From (7.45), it follows that $(1/\bar{\lambda})\overline{\varphi_j(t)}$, $j = 1, 2, \ldots, n$, are the Fourier coefficients of $\overline{k(t, s)}$—as function of s—with respect to the system $\{\varphi_j\}$, $j = 1, 2, \ldots, n$. We can apply the Bessel inequality, which yields

$$\frac{1}{|\lambda|^2} \sum_{j=1}^{n} |\varphi_j(t)|^2 \leq \int_a^b |k(t, s)|^2\, ds. \tag{7.46}$$

Integrating (7.46) and taking into account that $\{\varphi_j\}$ is an orthonormal system, we get

$$\frac{n}{|\lambda|^2} \leq \int_a^b \int_a^b |k(t, s)|^2\, dt\, ds \tag{7.47}$$

that is, n is bounded. Theorem 7.6 is completely proved.

Remark. Theorem 7.6 gives an answer to question 1, raised at the end of §7.2. It doesn't give the precise value of the dimension of the space of eigenfunctions, but only an upper estimate for this number.

To continue our discussion, we need the concept of *adjoint equation* or, which is the same, that of *adjoint kernel*. The *adjoint kernel* $k^*(t, s)$ of the kernel $k(t, s)$ is given by

$$k^*(t, s) = \overline{k(s, t)}, \qquad (t, s) \in [a, b] \times [a, b]. \tag{7.48}$$

Accordingly, the *adjoint equation* of (7.1) is

$$y(t) = g(t) + \lambda \int_a^b k^*(t, s)y(s)\, ds \tag{7.49}$$

and the corresponding homogeneous equation is

$$y(t) = \lambda \int_a^b k^*(t, s)y(s)\, ds. \tag{7.50}$$

Let us denote by $D^*(\lambda)$, $D^*(t, s; \lambda)$, and so on, the corresponding quantities for the kernel $k^*(t, s)$. It can be easily seen that

$$D^*(\bar{\lambda}) = \overline{D(\lambda)}, \qquad D^*(t, s; \bar{\lambda}) = \overline{D(s, t; \lambda)}. \tag{7.51}$$

Indeed, we have

$$k^*\begin{pmatrix} t_1 & \cdots & t_n \\ s_1 & \cdots & s_n \end{pmatrix} = \overline{k\begin{pmatrix} s_1 & \cdots & s_n \\ t_1 & \cdots & t_n \end{pmatrix}}$$

from which it follows that $d_n^* = \overline{d_n}$, $d_n^*(t, s) = \overline{d_n(s, t)}$. From the last relations, we obtain at once (7.51). From (7.51) we can derive the following statement: λ is an eigenvalue of (7.31) if and only if $\bar{\lambda}$ is an eigenvalue of (7.50).

Theorem 7.7

If λ is an eigenvalue of multiplicity n for the kernel $k(t, s)$, then $\bar{\lambda}$ is an eigenvalue of multiplicity n for the kernel $k^*(t, s)$.

Proof. Let $\{\varphi_j(t)\}$, $j = 1, 2, \ldots, m$, be an orthonormal basis of the space of eigenfunctions of the kernel $k(t, s)$, corresponding to λ:

$$\varphi_j(t) = \lambda \int_a^b k(t, s)\varphi_j(s)\, ds, \qquad j = 1, 2, \ldots, m. \tag{7.52}$$

Consider also an orthonormal basis $\{\psi_j(t)\}$, $j = 1, 2, \ldots, n$, of the space of eigenfunctions of $k^*(t, s)$, corresponding to $\bar{\lambda}$:

$$\psi_j(t) = \bar{\lambda} \int_a^b \overline{k(s, t)}\psi_j(s)\, ds, \qquad j = 1, 2, \ldots, n. \tag{7.53}$$

Assume, for instance, that $m < n$, and consider the kernel

$$K(t, s) = k(t, s) - \sum_{j=1}^{m} \overline{\varphi_j(s)}\psi_j(t). \tag{7.54}$$

The homogeneous adjoint equations

$$\varphi(t) = \lambda \int_a^b K(t, s)\varphi(s)\, ds \tag{7.55}$$

and

$$\psi(t) = \bar{\lambda} \int_a^b \overline{K(s, t)}\psi(s)\, ds \tag{7.56}$$

should have simultaneously either only the zero solution or nonzero solutions [because if λ is an eigenvalue for $K(t, s)$, $\bar{\lambda}$ is an eigenvalue for $\overline{K(s, t)}$].

We shall prove now that (7.54) has only the zero solution and (7.56) is verified by $\psi(t) = \psi_k(t)$, $k > m$. This would mean that the assumption $m < n$ cannot be accepted—that is, $m \geq n$. Inasmuch as the adjoint kernel of $k^*(t, s)$ is the kernel $k(t, s)$ itself, we shall have $n \geq m$—that is, $m = n$. Equations (7.55) and (7.56) can be written in the form

$$\varphi(t) = \lambda \int_a^b k(t, s)\varphi(s)\, ds - \lambda \sum_{j=1}^{m} \psi_j(t) \int_a^b \varphi(s)\overline{\varphi_j(s)}\, ds \tag{7.57}$$

$$\psi(t) = \bar{\lambda} \int_a^b \overline{k(s, t)}\psi(s)\, ds - \bar{\lambda} \sum_{j=1}^{m} \varphi_j(t) \int_a^b \psi(s)\overline{\psi_j(s)}\, ds. \tag{7.58}$$

We shall now multiply both sides of (7.57) by $\overline{\psi_k(t)}$, $k = 1, 2, \ldots, m$ and integrate from a to b. We get

$$\int_a^b \varphi(t)\overline{\psi_k(t)}\, dt = \int_a^b \left\{\lambda \int_a^b k(t, s)\overline{\psi_k(t)}\, dt\right\}\varphi(s)\, ds$$
$$- \lambda \sum_{j=1}^{m} \left(\int_a^b \psi_j(t)\overline{\psi_k(t)}\, dt\right) \cdot \left(\int_a^b \varphi(s)\overline{\varphi_j(s)}\, ds\right).$$

The quantity in braces from the first term of the right-hand side can be replaced by $\overline{\psi_k(s)}$, which yields

$$\int_a^b \varphi(s)\overline{\varphi_k(s)}\, ds = 0, \qquad k = 1, 2, \ldots, m. \tag{7.59}$$

Taking (7.59) into account, we reduce Eq. (7.57) to

$$\varphi(t) = \lambda \int_a^b k(t, s)\varphi(s)\, ds$$

which means that $\varphi(t)$ is an eigenfunction of $k(t, s)$, corresponding to λ.

Consequently, $\varphi(t) = \sum_{j=1}^{m} c_j \varphi_j(t)$, for some constants $c_j, j = 1, 2, \ldots, m$. But (7.59) shows that all c_j are zero because $c_j = (\varphi, \varphi_j), j = 1, 2, \ldots, m$. It follows that Eq. (7.55) has only the zero solution.

On the other hand, Eq. (7.56), or, which is the same, Eq. (7.58), is satisfied by any $\psi_k(t)$, $k > m$. Indeed, it reduces to

$$\psi_k(t) = \bar{\lambda} \int_a^b \overline{k(s, t)} \psi_k(s)\, ds, \qquad k > m$$

because $(\psi_k, \psi_j) = 0$ for $k > m$, $j = 1, 2, \ldots, m$. The last equation is nothing else but (7.53), for $k = m + 1, \ldots, n$. The proof of Theorem 7.7 is now complete.

Now we shall answer question 2 from the end of §7.2.

Theorem 7.8

Assume that λ is an eigenvalue of the kernel $k(t, s)$. Then Eq. (7.1) *has solutions if and only if $f(t)$ is orthogonal to the space of eigenfunctions of the adjoint kernel corresponding to $\bar{\lambda}$—that is, $(f, \psi) = 0$ for any solution of the equation*

$$\psi(t) = \bar{\lambda} \int_a^b \overline{k(s, t)} \psi(s)\, ds. \tag{7.60}$$

Proof. Let us assume that $x(t)$ is a solution of (7.1). Multiplying both sides of (7.1) by $\overline{\psi(t)}$—a solution of (7.60)—and integrating from a to b, we obtain

$$\int_a^b x(t)\overline{\psi(t)}\, dt = \int_a^b f(t)\overline{\psi(t)}\, dt + \int_a^b \left\{\lambda \int_a^b k(t, s)\overline{\psi(t)}\, dt\right\} x(s)\, ds.$$

The quantity in the braces can be replaced by $\overline{\psi(s)}$, which yields

$$\int_a^b f(t)\overline{\psi(t)}\, dt = (f, \psi) = 0. \tag{7.61}$$

Consequently, condition (7.61) is necessary for the existence of at least one solution of Eq. (7.1).

Let us prove now that (7.61) is also a sufficient condition for the existence of solutions of (7.1). Consider again the kernel $K(t, s)$ given by (7.54). We know that $m = n$. The equation

$$y(t) = f(t) + \lambda \int_a^b K(t, s) y(s)\, ds \tag{7.62}$$

has a unique solution for any continuous $f(t)$, inasmuch as Eq. (7.55) has only the zero solution (see the proof of Theorem 7.7). But Eq. (7.62) can be also written

$$y(t) = f(t) + \lambda \int_a^b k(t, s)y(s)\,ds - \sum_{j=1}^{n} \psi_j(t) \int_a^b y(s)\overline{\varphi_j(s)}\,ds. \tag{7.63}$$

Multiplying both sides of (7.63) by $\overline{\psi_k(t)}$ and integrating from a to b, we find

$$\int_a^b y(t)\overline{\psi_k(t)}\,dt = \int_a^b f(t)\overline{\psi_k(t)}\,dt + \int_a^b y(s)\left\{\lambda \int_a^b k(t, s)\overline{\psi_k(t)}\,dt\right\} ds - \lambda \int_a^b y(s)\overline{\varphi_k(s)}\,ds.$$

Replacing the term in the braces by $\overline{\psi_k(s)}$ and taking into account (7.61), we get $(y, \varphi_k) = 0$, $k = 1, 2, \ldots, n$. But these conditions show that (7.63) reduces to (7.1). Consequently, there exists at least one solution of (7.1), if f satisfies (7.61). The theorem is completely proved.

Remark. Obviously, Eq. (7.1) has, under condition (7.61), an infinity of solutions. They are given by

$$x(t) = \sum_{j=1}^{n} c_j \varphi_j(t) + y(t) \tag{7.64}$$

where $y(t)$ is a solution of (7.1). For instance, $y(t)$ could be the solution the existence of which was just proved and $\{\varphi_j(t)\}$, $j = 1, 2, \ldots, n$, form a basis of the space of eigenfunctions corresponding to λ.

Finally, let us summarize the discussion of Eq. (7.1), as developed in this chapter.

Consider Eq. (7.1) *with continuous kernel* $k(t, s)$ *on* $[a, b] \times [a, b]$ *and continuous* $f(t)$, λ *being a complex parameter. Then, either Eq.* (7.1) *has a unique solution for any* $f(t)$ *and the homogeneous equation* (7.31) *has only the zero solution, or Eq.* (7.31) *has nonzero solutions.*

In the last case, there exists only a finite number of linearly independent solutions of Eq. (7.31). *The adjoint homogeneous equation* (7.60) *has the same maximum number of linearly independent solutions as Eq.* (7.31). *Equation* (7.1) *has solutions if and only if* $f(t)$ *is orthogonal to any solution of the Eq.* (7.60).

PROBLEMS

1. Consider the homogeneous integral equation

$$x(t) = \lambda \int_0^{\pi} [\sin t \cos s] x(s)\,ds.$$

Prove that for any λ the only solution is $x(t) = 0$. In other words, the kernel $k(t, s) = \sin t \cos s$, $t, s \in [0, \pi]$, has no eigenvalues.

2. Assume that the kernel $k(t, s)$ is such that $k(t, s) = 0$ for $s > t$. Then, the corresponding integral equation reduces to an equation of Volterra type

$$x(t) = f(t) + \lambda \int_a^t k(t, s)x(s)\, ds, \qquad t \in [a, b].$$

Prove the existence of an entire function (with respect to λ) $R(t, s; \lambda)$, such that

$$x(t) = f(t) + \lambda \int_a^t R(t, s; \lambda) f(s)\, ds$$

for any continuous $f(t)$. What conclusion can we derive concerning the eigenvalues of a Volterra kernel?

3. Consider the integral equation with *degenerate* kernel

$$x(t) = f(t) + \lambda \int_a^b \left\{ \sum_{j=1}^n a_j(t) b_j(s) \right\} x(s)\, ds.$$

Notice that any solution should have the form

$$x(t) = f(t) + \lambda \sum_{j=1}^n C_j a_j(t)$$

where C_j $(j = 1, 2, \ldots, n)$ are constants to be determined. Without loss of generality, we can assume that $a_j(t)$ $(j = 1, 2, \ldots, n)$ are linearly independent on $[a, b]$. Prove that C_j satisfy an algebraic linear system. Discuss the existence problem and prove the main results of this chapter in the preceding particular case.

4. Consider the integral equation

$$x(t) = f(t) + \lambda \int_0^1 P(t, s)x(s)\, ds$$

where $P(t, s)$ is a homogeneous polynomial of degree n in t, s. Discuss the preceding equation.

5. Consider the singular integral equation

$$x(t) = \int_{-\infty}^{\infty} \exp(-|t - s|)x(s)\, ds.$$

Let α be a real number. Prove that $x(t) = \exp(i\alpha t)$ is a solution of the preceding equation, corresponding to $\lambda = (1 + \alpha^2)/2$. Therefore, the set of eigenvalues of a singular integral equation is not necessarily countable.

6. Let a be a positive number. Then

$$\int_0^\infty e^{-as} \sin ts\, ds = \frac{t}{a^2 + t^2}$$

and

$$\int_0^\infty \frac{s \sin ts}{a^2 + s^2}\, ds = \left(\frac{\pi}{2}\right)^{1/2} e^{-at}.$$

Obviously, the integral equation

$$x(t) = \lambda \int_0^\infty \sin(ts)x(s)\, ds$$

has infinitely many eigenfunctions corresponding to the eigenvalue $(2/\pi)^{1/2}$. Namely, these eigenfunctions are

$$\frac{t}{a^2+t^2} + \left(\frac{\pi}{2}\right)^{1/2} e^{-at}, \qquad a > 0.$$

They are linearly independent. Of course, this situation, which apparently contradicts the result given by Theorem 7.6, occurs because the integral equation under consideration is singular (more precisely, the interval of integration has infinite measure).

7. Investigate the integral equation with singular kernel

$$x(t) = \lambda \int_a^b \frac{k(t, s)}{(t-s)^\alpha} x(s)\, ds$$

where α is such that $0 < \alpha < 1$ and $k(t, s)$ is continuous. Prove that the iterated kernels $k_n(t, s)$ are bounded for sufficiently large n. Use this property in solving the nonhomogeneous integral equation associated with the equation considered above.

8. Consider the nonlinear Fredholm equation

$$x(t) = f(t) + \int_a^b k(t, s, x(s))\, ds.$$

Prove the existence and uniqueness of the solution under the main assumption that $k(t, s, x)$ satisfies a Lipschitz condition with respect to x, the Lipschitz constant being sufficiently small. An alternate condition is that the length of the interval $[a, b]$ is small enough. Formulate the exact conditions.

9. A noteworthy particular case of the equation considered in problem 8 is the so-called Hammerstein integral equation

$$x(t) = f(t) + \int_a^b k(t, s)g(s, x(s))\, ds.$$

Discuss the existence problem and give sufficient conditions assuring both existence and uniqueness of the solution.

10. Let us consider the integro-differential equation

$$\dot{x}(t) = f(t) + \int_a^b k(t, s, x(s))\, ds.$$

Find conditions under which there exists at least one solution satisfying an initial condition $x(a) = x_0$.

11. Discuss the following extensions of the results established in this chapter:
(a) $x(t)$ is a vector function with n coordinates;
(b) The unknown function depends on several independent variables.

8

Theory of Self-Adjoint Integral Equations and Some Applications

If the kernel of an integral equation is such that $k(t, s) = \overline{k(s, t)}$, then the adjoint equation coincides with the given one. Such kernels are called *Hermitian kernels.* We have seen that there exist continuous kernels without any eigenvalue (see problem 1, Chapter 7). Among other interesting properties, the Hermitian kernels possess that of the existence of at least one eigenvalue.

We owe to Hilbert the main results of the theory of integral equations with Hermitian kernel. Of particular interest are those results of this theory which are related to the series expansion of some functions. Theorem 8.2 provides a powerful criterion of convergence, with many useful applications.

§8.1. Some Properties of an Integral Operator

If $k(t, s)$ is a continuous kernel on $[a, b] \times [a, b]$, then it defines an *integral operator*, which we shall denote by K, given by

$$(Kx)(t) = \int_a^b k(t, s)x(s)\, ds, \qquad x \in C[a, b]. \tag{8.1}$$

Formula (8.1) defines for any continuous function x another continuous function Kx. A similar case was encountered in §6.4. It is easy to see that

K is a linear homogeneous operator, which means that $K(x + y) = Kx + Ky$ and $K(\alpha x) = \alpha Kx$, where $x, y \in C[a, b]$ and α is a complex number.

Another interesting property is the *boundedness* of K—that is, there exists a positive number M, such that

$$\|Kx\| \leq M \|x\|, \qquad x \in C[a, b] \tag{8.2}$$

where $\|\cdot\|$ denotes the norm we defined in §7.3 [see formula (7.35)]. The number M can be chosen as follows:

$$M^2 = \int_a^b \int_a^b |k(t, s)|^2 \, dt \, ds. \tag{8.3}$$

Indeed, for any fixed $t \in [a, b]$, we can apply the Schwarz inequality and find

$$|(Kx)(t)|^2 \leq \left\{\int_a^b |k(t, s)|^2 \, ds\right\} \int_a^b |x(s)|^2 \, ds, \qquad t \in [a, b]. \tag{8.4}$$

Integrating both sides of (8.4) from a to b, we obtain

$$\|Kx\|^2 \leq \left(\int_a^b \int_a^b |k(t, s)|^2 \, dt \, ds\right) \|x\|^2 \tag{8.5}$$

which yields (8.2), if M is given by (8.3).

Remark. The least number M with the property (8.2) is called the *norm of the operator* K and is denoted by $\|K\|$. It can be defined also by

$$\|K\| = \sup_{\|x\| = 1} \|Kx\|. \tag{8.6}$$

Indeed, from (8.6) it follows that $\|Kx\| \leq \|K\|$ for any $x \in C[a, b]$, such that $\|x\| = 1$. If $y \in C[a, b]$, $y \neq 0$, then $y/\|y\|$ has the norm equal to 1, which yields $\|Ky\| \leq \|K\| \, \|y\|$. On the other hand, from (8.6) it follows also that $\|K\|$ is the least number with the required property.

We shall now establish a result concerning the norm of an integral operator corresponding to a Hermitian kernel.

Lemma

Let $k(t, s)$ be a Hermitian kernel. Consider the operator defined by (8.1). *Then*

$$\|K\| = \sup_{\|x\| = 1} |(Kx, x)|. \tag{8.7}$$

Proof. Inasmuch as $|(Kx, x)| \leq \|Kx\| \, \|x\| \leq \|K\| \, \|x\|^2 = \|K\|$ for $\|x\| = 1$,

$$\sup_{\|x\| = 1} |(Kx, x)| \leq \|K\|. \tag{8.8}$$

We have to prove the opposite inequality

$$\sup_{\|x\|=1} |(Kx, x)| \geq \|K\|. \tag{8.9}$$

Denoting by A the left-hand side of (8.9), we can write

$$|(Kx, x)| \leq A\|x\|^2, \qquad x \in C[a, b]. \tag{8.10}$$

Indeed, we can divide both sides of (8.10) by $\|x\|^2$ and obtain $|(Ky, y)| \leq A$, with $\|y\| = 1$ $(y = x/\|x\|)$. Inasmuch as

$$(Kx, y) = \int_a^b \int_a^b k(t, s)x(s)\overline{y(t)}\, dt\, ds$$

it is easy to see that $(Kx, y) = (x, Ky) = \overline{(Ky, x)}$, taking into account that $k(t, s)$ is a Hermitian kernel. In particular, $(Kx, x) = (x, Kx) = \overline{(Kx, x)}$, which means that (Kx, x) is a real number.

Consider now two arbitrary functions x and y, with norm equal to unity. Let us remark that

$$(K(x + y), x + y) = (Kx, x) + (Ky, y) + 2\,\mathrm{Re}(Kx, y) \leq A\|x + y\|^2$$

and

$$(K(x - y), x - y) = (Kx, x) + (Ky, y) - 2\,\mathrm{Re}(Kx, y) \geq -A\|x - y\|^2.$$

By subtraction, we get

$$4\,\mathrm{Re}(Kx, y) \leq A(\|x + y\|^2 + \|x - y\|^2) = 2A(\|x\|^2 + \|y\|^2) = 4A$$

which means that

$$\mathrm{Re}(Kx, y) \leq A, \qquad \|x\| = \|y\| = 1. \tag{8.11}$$

If we choose $y = Kx/\|Kx\|$, for $\|Kx\| \neq 0$, inequality (8.11) yields

$$\|Kx\| \leq A \quad \text{for} \quad \|x\| = 1. \tag{8.12}$$

For $Kx = 0$, (8.12) is obviously verified. Taking the supremum of the left-hand side in (8.12), for $\|x\| = 1$, we obtain (8.9). The lemma is proved.

§8.2. The Existence of Eigenvalues

Consider the homogeneous equation

$$x(t) = \lambda \int_a^b k(t, s)x(s)\, ds \tag{8.13}$$

where $k(t, s)$ is a Hermitian kernel that does not vanish identically in $[a, b] \times [a, b]$, and prove that it has at least one eigenvalue. We would like to point out that the proof of this result has a constructive character.

Theorem 8.1

If $k(t, s)$ is a Hermitian kernel not vanishing identically, then either $\|K\|^{-1}$ or $-\|K\|^{-1}$ is an eigenvalue of Eq. (8.13).

Proof. First, $\|K\| > 0$. Indeed, from $k(t, s) \not\equiv 0$ in $[a, b] \times [a, b]$, it follows that there exists at least one $t_0 \in [a, b]$ such that $k(t_0, s) \not\equiv 0$ in $[a, b]$. For $x(s) = k(s, t_0)$, we get

$$(Kx)(t_0) = \int_a^b |k(t_0, s)|^2 \, ds > 0$$

which implies $\|Kx\| > 0$. This inequality can hold only if $\|K\| > 0$ because $\|Kx\| \leq \|K\| \cdot \|x\|$.

The lemma we proved in §8.1 shows that (8.7) holds. This fact implies the existence of a sequence $\{x_n(t)\}$ of continuous functions on $[a, b]$, such that $\|x_n\| = 1$, $n = 1, 2 \ldots$, and

$$\lim_{n \to \infty} (Kx_n, x_n) = \pm \|K\| . \tag{8.14}$$

Assume that (8.14) holds, the plus sign being valid in the right-hand side, and denote, as previously, $A = \|K\|$.

The sequence of continuous functions $\{(Kx_n)(t)\}$ is uniformly bounded and equicontinuous on $[a, b]$. Indeed, we have

$$|(Kx_n)(t)|^2 \leq \left\{\int_a^b |k(t, s)|^2 \, ds\right\} \int_a^b |x_n(s)|^2 \, ds = \int_a^b |k(t, s)|^2 \, ds$$

and

$$|(Kx_n)(t) - (Kx_n)(u)|^2 \leq \int_a^b |k(t, s) - k(u, s)|^2 \, ds$$

from which the assertion follows immediately, because $k(t, s)$ is a continuous function on $[a, b] \times [a, b]$.

Consequently, without any loss of generality, we can assume that the sequence $\{(Kx_n)(t)\}$ is uniformly convergent on $[a, b]$. Let us denote $x(t) = \lim_{n \to \infty} (Kx_n)(t)$ and show that $x(t) \not\equiv 0$. Indeed,

$$\|Kx_n - Ax_n\|^2 = (Kx_n - Ax_n, Kx_n - Ax_n) = \|Kx_n\|^2 + A^2 - 2A(Kx_n, x_n) \tag{8.15}$$

because $\|x_n\| = 1$ and $(x_n, Kx_n) = \overline{(Kx_n, x_n)} = (Kx_n, x_n)$. Inasmuch as

$\{Kx_n\}$ is uniformly convergent, $\|Kx_n\|^2 \to \|x\|^2$ as $n \to \infty$, and, taking into account (8.14), we find

$$\lim_{n\to\infty} \|Kx_n - Ax_n\|^2 = \|x\|^2 + A^2 - 2A^2 = \|x\|^2 - A^2 \geq 0$$

which means that $\|x\| \geq A > 0$. This implies $x(t) \not\equiv 0$ on $[a, b]$.

On the other hand, $\|Kx_n\| \leq A\|x_n\| = A$, from which we get, taking (8.15) into account,

$$\lim_{n\to\infty} \|Kx_n - Ax_n\|^2 \leq A^2 + A^2 - 2A^2 = 0$$

that is,

$$\lim_{n\to\infty} \|Kx_n - Ax_n\|^2 = 0.$$

Inasmuch as $\|Ky\| \leq A\|y\|$ for any continuous y, we shall have

$$\|K(Kx_n - Ax_n)\| \leq A\|Kx_n - Ax_n\| \to 0$$

as $n \to \infty$. We can write

$$K(Kx_n - Ax_n) = K(Kx_n) - AKx_n$$

and from $Kx_n \to x$, uniformly on $[a, b]$, we get $\|Kx - Ax\| = 0$—that is, $Kx - Ax = 0$. This can be written as

$$\int_a^b k(t, s)x(s)\, ds = Ax(t), \qquad t \in [a, b] \tag{8.16}$$

which shows that $x(t)$ is an eigenfunction of the kernel $k(t, s)$, corresponding to the eigenvalue A^{-1}.

If the minus sign holds in (8.14), then we have to start from $\|Kx_n + Ax_n\|^2 \geq 0$ and develop a similar argument. We get that $-A^{-1}$ is an eigenvalue. Theorem 8.1 is thus proved.

Before we proceed further, it is convenient to present two fundamental properties of the Hermitian kernels, which follow easily from the previous considerations.

1. The eigenvalues of a Hermitian kernel are real.
2. If x and y are eigenfunctions of a Hermitian kernel, corresponding to distinct eigenvalues, then $(x, y) = 0$.

Concerning the first property, we have to remark that $x = \lambda Kx$ implies $(x, x) = \lambda(Kx, x)$. Inasmuch as (Kx, x) is a real number, as proved, and $(x, x) > 0$, it follows that λ is real.

Assume now $x = \lambda Kx$, $y = \mu Ky$, with $\lambda \neq \mu$. We have $(x, y) = \lambda(Kx, y) = \lambda(x, Ky) = \lambda\mu^{-1}(x, y)$. Because $\lambda\mu^{-1} \neq 1$, the preceding relation implies $(x, y) = 0$.

§8.3. The Hilbert–Schmidt Expansion Theorem

Assume that the Hermitian kernel $k(t, s)$ has infinitely many eigenvalues $\{\lambda_n\}$, $n = 1, 2, \ldots$. We know that they form a sequence having as unique limit the point at infinity (because they are the zeros of an entire function). It is possible to range these eigenvalues in an ordered sequence—namely, such that

$$|\lambda_1| \leq |\lambda_2| \leq \cdots \leq |\lambda_n| \leq \cdots. \tag{8.17}$$

We agree to take into account the multiplicity of the eigenvalues—that is, if λ_j is of multiplicity p, then it will appear p times in the sequence (8.17).

We have seen that any finite-dimensional vector space has an orthonormal basis. We agree also to consider in every space of eigenfunctions, corresponding to a given eigenvalue of $k(t, s)$, an orthonormal basis. Inasmuch as the eigenfunctions corresponding to distinct eigenvalues are orthogonal, we can conclude that the set of eigenfunctions of the kernel $k(t, s)$ form an orthogonal sequence $\{\varphi_n\}$, $n = 1, 2, \ldots$. There exists a one-to-one mapping between the set of eigenvalues, as described, and the set of eigenfunctions.

The expansion theorem we intend to prove can be formulated as follows.

Theorem 8.2

Assume that $k(t, s)$ is a Hermitian kernel with infinitely many eigenvalues. If the function $g(t)$ is such that

$$g(t) = (Kf)(t) = \int_a^b k(t, s)f(s)\, ds \tag{8.18}$$

with $f \in C[a, b]$, then

$$g(t) = \sum_{n=1}^{\infty} (g, \varphi_n)\varphi_n(t), \qquad t \in [a, b] \tag{8.19}$$

the series in the right-hand side of (8.19) *being uniformly and absolutely convergent.*

Proof. First, let us compute the Fourier coefficients (g, φ_n). We have

$$(g, \varphi_n) = (Kf, \varphi_n) = (f, K\varphi_n) = \lambda_n^{-1}(f, \varphi_n)$$

because $\varphi_n = \lambda_n K\varphi_n$, $n = 1, 2, \ldots$. From $\lambda_n^{-1}\varphi_n(t) = \int_a^b k(t, s)\varphi_n(s)\, ds$, it follows that $\lambda_n^{-1}\overline{\varphi_n(t)}$ are the Fourier coefficients of $\overline{k(t, s)}$ as a function of s, with respect to $\{\varphi_n\}$. Applying the inequality of Bessel, we find

$$\sum_{j=m}^{n} \lambda_j^{-2} |\varphi_j(t)|^2 \le \int_a^b |k(t, s)|^2 \, ds \le B^2, \qquad t \in [a, b]. \tag{8.20}$$

On the other hand,

$$\sum_{j=m}^{n} |(g, \varphi_j)| \, |\varphi_j(t)| = \sum_{j=m}^{n} |(f, \varphi_j)| \cdot |\lambda_j^{-1} \varphi_j(t)|$$

$$\le \left\{ \sum_{j=m}^{n} |(f, \varphi_j)|^2 \right\}^{1/2} \cdot \left\{ \sum_{j=m}^{n} \lambda_j^{-2} |\varphi_j(t)|^2 \right\}^{1/2} \le B \left\{ \sum_{j=m}^{n} |(f, \varphi_j)|^2 \right\}^{1/2}$$

We know that $\sum_{j=1}^{\infty} |(f, \varphi_j)|^2 \le \|f\|^2$, which implies

$$\sum_{j=m}^{n} |(f, \varphi_j)|^2 < B^{-1}\varepsilon, \quad \text{for} \quad m, n \ge N(\varepsilon).$$

Combining the above inequalities, we get

$$\sum_{j=m}^{n} |(g, \varphi_j)| \cdot |\varphi_j(t)| < \varepsilon, \quad \text{for} \quad m, n \ge N(\varepsilon) \tag{8.21}$$

which proves that series (8.19) is uniformly and absolutely convergent on $[a, b]$.

We have to prove now that the series is convergent to $g(t)$—that is, that (8.19) holds. Let us consider the auxiliary kernels

$$k_n(t, s) = k(t, s) - \sum_{j=1}^{n} \lambda_j^{-1} \varphi_j(t)\overline{\varphi_j(s)}. \tag{8.22}$$

Because λ_j are real numbers, we get $k_n(s, t) = \overline{k_n(t, s)}$. Moreover, we have

$$K_n f = Kf - \sum_{j=1}^{n} \lambda_j^{-1}(f, \varphi_j)\varphi_j = Kf - \sum_{j=1}^{n} (f, \lambda_j^{-1}\varphi_j)\varphi_j$$

$$= Kf - \sum_{j=1}^{n} (f, K\varphi_j)\varphi_j = Kf - \sum_{j=1}^{n} (Kf, \varphi_j)\varphi_j = g - \sum_{j=1}^{n} (g, \varphi_j)\varphi_j.$$

Consequently, it suffices to prove that

$$\lim_{n \to \infty} (K_n f)(t) = 0 \tag{8.23}$$

in the sense of the convergence in the mean.

Indeed, from (8.22) we have

$$(K_n f, \varphi_m) = (Kf, \varphi_m) - (Kf, \varphi_m) = 0 \quad \text{for} \quad m \le n. \tag{8.24}$$

Thus, any eigenfunction of the kernel $k_n(t, s)$ is orthogonal to φ_m, $m \le n$. Indeed, if f is an eigenfunction corresponding to δ, then $K_n f = \delta^{-1} f$ and (8.24) becomes $(f, \varphi_m) = 0$.

Now, assume that δ is an eigenvalue of $k_n(t, s)$, and prove that it is also an eigenvalue of $k(t, s)$. Inasmuch as $(f, \varphi_m) = 0$ for $m \leq n$ [see (8.24)], it follows that $K_n f = Kf$. Hence, $Kf = \delta^{-1} f$, which shows that δ is an eigenvalue of $k(t, s)$. Moreover, because $(f, \varphi_m) = 0$, $m \leq n$, δ cannot be among the eigenvalues λ_m, $m \leq n$. This leads to $|\delta| \geq |\lambda_n|$.

On the other hand, we know that at least one of the numbers $\pm \|K_n\|^{-1}$ is an eigenvalue for $k_n(t, s)$ (see Theorem 8.1). Consequently,

$$\|K_n\| \leq |\lambda_n|^{-1} \tag{8.25}$$

which implies $\lim_{n\to\infty} \|K_n\| = 0$. Inasmuch as $\|K_n f\| \leq \|K_n\| \cdot \|f\|$, (8.25) implies $\lim_{n\to\infty} \|K_n f\| = 0$, for any continuous f. In other words, the sequence $\{K_n f\}$ is convergent to zero in the mean.

But we already proved that $\{K_n f\}$ is uniformly convergent on $[a, b]$:

$$(K_n f)(t) = g(t) - \sum_{j=1}^{n} \lambda_j^{-1}(f, \varphi_j) \cdot \varphi_j(t) \to g(t) - \sum_{j=1}^{\infty} \lambda_j^{-1}(f, \varphi_j)\varphi_j(t).$$

Hence, $\|\lim_{n\to\infty} (K_n f)(t)\| = 0$, because we can permute the integral and the limit, the convergence being uniform on $[a, b]$. We can write

$$\left\| g(t) - \sum_{j=1}^{\infty} \lambda_j^{-1}(f, \varphi_j)\varphi_j(t) \right\| = 0$$

and because the function inside is continuous, we get (8.19). The theorem is thus proved.

An immediate application of Theorem 8.2 is related to the nonhomogeneous equation

$$x(t) = f(t) + \lambda \int_a^b k(t, s)x(s)\, ds \tag{8.26}$$

where $k(t, s)$ is a Hermitian kernel. We shall find the series expansion of the solution. Inasmuch as

$$x(t) - f(t) = \lambda \int_a^b k(t, s)x(s)\, ds$$

it follows that $x(t) - f(t) = g(t)$ satisfies the condition required by the Hilbert–Schmidt theorem.

Assume that λ is not an eigenvalue of $k(t, s)$. Then $(g, \varphi_j) = \lambda\lambda_j^{-1}(x, \varphi_j) = \lambda\lambda_j^{-1}(f + g, \varphi_j)$. This leads to

$$(g, \varphi_j) = \frac{\lambda}{\lambda_j - \lambda}(f, \varphi_j), \qquad j = 1, 2, \ldots \tag{8.27}$$

and, consequently,

$$x(t) = f(t) + \lambda \sum_{j=1}^{\infty} \frac{(f, \varphi_j)}{\lambda_j - \lambda} \varphi_j(t). \tag{8.28}$$

If λ coincides with λ_n, then (8.27) has no meaning for those j such that $\lambda_j = \lambda_n$. In this case, (8.26) has solutions if and only if $(f, \varphi_j) = 0$ for all j for which $\lambda_j = \lambda_n$. The expansion is now

$$x(t) = f(t) + \lambda \sum_{\lambda_j \neq \lambda_n} \frac{(f, \varphi_j)}{\lambda_j - \lambda} \varphi_j(t) + \sum_{\lambda_j = \lambda_n} c_j \varphi_j(t) \tag{8.29}$$

where c_j are arbitrary constants (their number is the same as the multiplicity of λ_n).

Both series (8.28) and (8.29) are uniformly and absolutely convergent on $[a, b]$. Hence, if we know the eigenvalues and the eigenfunctions of a Hermitian kernel $k(t, s)$, it is possible to represent the solutions of Eq. (8.26) by means of some series in terms of eigenfunctions.

§8.4. Complete Kernels and Systems

The concepts of complete kernel and complete system of functions are necessary in the study of some integral equations we shall consider in the next section, in connection with Sturm–Liouville problems (boundary value problems) for second-order linear equations depending on a parameter. First, let us prove the following orthogonality result concerning the real Hermitian kernels [i.e., symmetric: $k(t, s) = k(s, t)$].

Lemma

If $k(t, s)$ is a continuous symmetric kernel with infinitely many eigenvalues, then conditions (8.30) *and* (8.31) *are equivalent*:

$$\int_a^b k(t, s)x(s)\, ds = 0, \qquad t \in [a, b] \tag{8.30}$$

$$\int_a^b \varphi_j(s)x(s)\, ds = 0, \qquad j = 1, 2, \ldots. \tag{8.31}$$

Here, $x(t) \in C[a, b]$ is a given function and $\{\varphi_j\}$ is the system of eigenfunctions of the kernel $k(t, s)$.

Proof. Assume (8.30) is true, multiply it by $\varphi_j(t)$, and integrate from a to b. We get

$$\int_a^b \varphi_j(t)\, dt \int_a^b k(t, s)x(s)\, ds = \int_a^b x(s)\, ds \int_a^b k(s, t)\varphi_j(t)\, dt$$

$$= \lambda_j^{-1} \int_a^b \varphi_j(s)x(s)\, ds = 0.$$

Hence, (8.30) implies (8.31). Conversely, assume (8.31) holds for all the eigenfunctions of $k(t, s)$. Consider the kernel

$$k_n(t, s) = k(t, s) - \sum_{j=1}^{n} \lambda_j^{-1} \varphi_j(t)\varphi_j(s)$$

[it is the same as (8.22), because $\varphi_j(t)$ are now real functions]. We obtain

$$\int_a^b k(t, s)x(s)\, ds = \int_a^b k_n(t, s)x(s)\, ds. \tag{8.32}$$

Therefore,

$$\left\| \int_a^b k(t, s)x(s)\, ds \right\| = \|K_n x\| \to 0 \quad \text{as} \quad n \to \infty$$

as shown in the proof of the expansion theorem (because $\|K_n x\| \leq \|K_n\| \cdot \|x\|$ and $\|K_n\| \to 0$ for $n \to \infty$). This means that $\| \int_a^b k(t, s)x(s)\, ds\| = 0$, which implies (8.30).

We shall say that the symmetric kernel $k(t, s)$ is *complete* if any relation of the form (8.30), with $x(t) \in C[a, b]$, implies $x(t) \equiv 0$. The orthonormal system $\{\varphi_j(t)\}$ is *complete* on $[a, b]$ if conditions (8.31), with $x(t) \in C[a, b]$, implies $x(t) \equiv 0$. It is almost trivial that any complete system contains infinitely many elements (the space of all continuous functions has no finite basis).

From the lemma proved previously, it follows that: *The symmetric kernel $k(t, s)$ is complete if and only if the system of its eigenvalues is complete.*

§8.5. The Sturm–Liouville Problem

One of the most interesting applications of Hilbert theory concerning the integral equation with symmetric kernel is related to the second-order differential equation

$$\frac{d}{dt}\left[p(t)\frac{dx}{dt}\right] + [\lambda - q(t)]x = 0 \tag{8.33}$$

with *boundary value conditions*:

$$\alpha_1 x(a) + \alpha_2 \dot{x}(a) = 0, \qquad \beta_1 x(b) + \beta_2 \dot{x}(b) = 0. \tag{8.34}$$

We shall always assume in the sequel that we are dealing with real quantities, and the following conditions hold:

1. $p(t)$ is continuously differentiable on $[a, b]$ and $p(t) > 0$ on the same interval;

2. $q(t)$ is a continuous function on $[a, b]$;
3. $|\alpha_1| + |\alpha_2| > 0$ and $|\beta_1| + |\beta_2| > 0$.

It is assumed that λ is a parameter.

The *Sturm–Liouville problem* consists in finding those values of λ for which there exists a nonzero solution of Eq. (8.33) defined on $[a, b]$, with boundary conditions (8.34). These values of λ are called *eigenvalues*, and the corresponding solutions are called *eigenfunctions*. We have seen that $\lambda = 0$ is never an eigenvalue of an integral equation. On the contrary, $\lambda = 0$ *could be an eigenvalue for the Sturm–Liouville problem* (see problem 3). The following two properties will be used further:

1. Any eigenvalue has the multiplicity equal to 1—that is, two distinct eigenfunctions corresponding to the same eigenvalue differ by a constant factor.
2. Two eigenfunctions corresponding to distinct eigenvalues are orthogonal.

Let us prove property 1. Assume that there exist two linearly independent eigenfunctions corresponding to the same eigenvalue λ—say, $x_1(t)$ and $x_2(t)$. Then $x(t) = C_1 x_1(t) + C_2 x_2(t)$ represents the general integral of (8.33) on $[a, b]$. It follows that any solution of (8.33) satisfies the boundary conditions (8.34). But this is impossible because we can choose $x(a)$ and $\dot{x}(a)$ as we like—consequently, such that the first condition (8.34) does not take place.

Concerning property 2, assume λ and μ are distinct eigenvalues and denote by $x(t)$, $y(t)$ the corresponding eigenfunctions. From

$$\frac{d}{dt}\left[p(t)\frac{dx}{dt}\right] + [\lambda - q(t)]x = 0,$$

$$\frac{d}{dt}\left[p(t)\frac{dy}{dt}\right] + [\mu - q(t)]y = 0$$

we easily get

$$\frac{d}{dt}\left[p(t)\left(y\frac{dx}{dt} - x\frac{dy}{dt}\right)\right] = (\mu - \lambda)xy. \tag{8.35}$$

Indeed, we have to multiply the equation for x by y and that for y by x. Subtracting, we obtain (8.35). Integrating both sides of (8.35) from a to b and taking into account

$$y(a)\dot{x}(a) - x(a)\dot{y}(a) = 0, \qquad y(b)\dot{x}(b) - x(b)\dot{y}(b) = 0 \tag{8.36}$$

we find

$$\int_a^b x(t)y(t)\,dt = 0. \tag{8.37}$$

Now, to prove (8.36), we shall start from (8.34), verified by both x and y. For instance, $\alpha_1 x(a) + \alpha_2 \dot{x}(a) = 0$, $\alpha_1 y(a) + \alpha_2 \dot{y}(a) = 0$, and $|\alpha_1| + |\alpha_2| > 0$ imply the first condition (8.36). A similar argument is applicable at $t = b$. The analogy between the theory of integral equations and that of the Sturm–Liouville problem is now obvious.

Actually, it is always possible to reduce the Sturm–Liouville problem to an integral equation with symmetric kernel. We shall be concerned with this reduction in the next two sections. There are two distinct cases we have to consider.

1. $\lambda = 0$ *is not an eigenvalue of the Sturm–Liouville problem*;
2. $\lambda = 0$ *is an eigenvalue.*

§8.6. Reduction to an Integral Equation: The First Case

Consider the differential equation

$$L(x) = \frac{d}{dt}\left[p(t)\frac{dx}{dt}\right] - q(t)x = 0 \tag{8.38}$$

which has no solution satisfying (8.34), other than $x(t) \equiv 0$. Therefore, $\lambda = 0$ is not an eigenvalue of the Sturm–Liouville problem (8.33), (8.34).

Let $x_1(t)$ be a solution of (8.38), satisfying the first condition of (8.34) and nonvanishing identically on $[a, b]$. By $x_2(t)$ we shall denote a solution of (8.38) such that the second condition of (8.34) is satisfied and $x_2(t) \not\equiv 0$. From Liouville's formula, we get

$$x_1(t)\dot{x}_2(t) - \dot{x}_1(t)x_2(t) = \frac{C}{p(t)}, \qquad t \in [a, b] \tag{8.39}$$

because Eq. (8.38) can be written

$$\ddot{x}(t) + \frac{p'(t)}{p(t)}\dot{x}(t) - \frac{q(t)}{p(t)}x(t) = 0.$$

We claim that $x_1(t)$ and $x_2(t)$ must be linearly independent, which means that $C \neq 0$. Otherwise both $x_1(t)$ and $x_2(t)$ would satisfy Eq. (8.38) and conditions (8.34), contrary to our assumption that $\lambda = 0$ is not an eigenvalue. Inasmuch as $x_1(t)$ and $x_2(t)$ can be chosen such that $C = -1$ (each one is determined up to a constant factor), we have a fundamental system of solutions such that

$$x_1(t)\dot{x}_2(t) - \dot{x}_1(t)x_2(t) = -\frac{1}{p(t)}, \qquad t \in [a, b]. \tag{8.40}$$

Consider now the nonhomogeneous equation corresponding to (8.38)

$$L(x) = -f(t) \tag{8.41}$$

where $f(t)$ is a continuous function on $[a, b]$. It is our aim to express the solution of (8.41), with boundary condition (8.34), using the method of variation of parameters. We shall seek the solution in the form

$$x(t) = C_1(t)x_1(t) + C_2(t)x_2(t) \tag{8.42}$$

where $C_1(t)$ and $C_2(t)$ must be found from the following system:

$$\begin{aligned} &\dot{C}_1(t)x_1(t) + \dot{C}_2(t)x_2(t) = 0 \\ &\dot{C}_1(t)\dot{x}_1(t) + \dot{C}_2(t)\dot{x}_2(t) = -\frac{f(t)}{p(t)}. \end{aligned} \tag{8.43}$$

From (8.43) and (8.40) we obtain

$$\dot{C}_1(t) = -x_2(t)f(t), \qquad \dot{C}_2(t) = x_1(t)f(t)$$

which implies

$$C_1(t) = \int_t^b x_2(s)f(s)\,ds, \qquad C_2(t) = \int_a^t x_1(s)f(s)\,ds \tag{8.44}$$

with convenient constants of integration. From (8.42) and (8.44) we get

$$x(t) = x_1(t)\int_t^b x_2(s)f(s)\,ds + x_2(t)\int_a^t x_1(s)f(s)\,ds \tag{8.45}$$

which can also be written as

$$x(t) = \int_a^b G(t, s)f(s)\,ds \tag{8.46}$$

with

$$G(t, s) = \begin{cases} x_1(s)x_2(t) & s \le t \\ x_1(t)x_2(s) & s > t. \end{cases} \tag{8.47}$$

The function $G(t, s)$ is obviously continuous on $[a, b] \times [a, b]$. It possesses also some differentiability properties that can be derived from (8.47). Function $G(t, s)$ is called the *Green's function*, associated with our Sturm–Liouville problem.

That (8.45)—or, equivalently, (8.46)—gives a solution of (8.41), is now obvious. We have to prove that $x(t)$ given by (8.45) satisfies both conditions (8.34). Indeed, from (8.45) we find $x(a) = x_1(a) \int_a^b x_2(s)f(s)\,ds$. Taking the derivative in both sides of (8.45), we obtain:

$$\dot{x}(t) = \dot{x}_1(t)\int_t^b x_2(s)f(s)\,ds + \dot{x}_2(t)\int_a^t x_1(s)f(s)\,ds$$

which yields $\dot{x}(a) = \dot{x}_1(a) \int_a^b x_2(s)f(s)\,ds$. Consequently,

$$\alpha_1 x(a) + \alpha_2 \dot{x}(a) = [\alpha_1 x_1(a) + \alpha_2 \dot{x}_1(a)] \int_a^b x_2(s)f(s)\,ds = 0$$

which means that $x(t)$ satisfies the first condition (8.34). A similar argument is valid for $t = b$—that is, $x(t)$ satisfies also the second condition (8.34).

Hence, we have proved that Eq. (8.41) has at least one solution satisfying (8.34). That this solution is unique can be derived from the fact that the difference of two solutions of (8.41) satisfies Eq. (8.38).

Now, the reduction of the Sturm–Liouville problem (8.33), (8.34) to an integral equation follows immediately from the previous considerations. Inasmuch as Eq. (8.33) can be written

$$L(x) = -\lambda x(t) \tag{8.48}$$

it follows that any solution of the Sturm–Liouville problem will satisfy the integral equation

$$x(t) = \lambda \int_a^b G(t, s)x(s)\,ds, \qquad t \in [a, b] \tag{8.49}$$

with $G(t, s)$ given by (8.47). Indeed, (8.49) can be obtained from (8.46) with $f(t) = \lambda x(t)$. Conversely, from (8.49) follows (8.48), which proves the equivalence of the Sturm–Liouville problem and of Eq. (8.49).

We can now apply the Hilbert theory for integral equations with symmetric kernel to Eq. (8.49), because $G(t, s) \not\equiv 0$. This last statement can be proved as follows: From (8.47) we get $G(t, t) = x_1(t)x_2(t)$, $t \in [a, b]$, and if $x_1(t_0) \neq 0$ with fixed $t_0 \in [a, b]$, we have $x_1(t) \neq 0$ for all t sufficiently near t_0. If $G(t, t)$ would be identically zero, then we should have $x_2(t) \equiv 0$ in a neighborhood of t_0. This implies $\dot{x}_2(t) \equiv 0$ in the same neighborhood, and the uniqueness shows that $x_2(t) \equiv 0$ on $[a, b]$. This contradicts the fact that $x_1(t)$ and $x_2(t)$ form a fundamental system for (8.38).

Moreover, *the kernel* $G(t, s)$ *is complete.* Indeed, if there exists a continuous function $f(t) \not\equiv 0$ on $[a, b]$ such that it is orthogonal to $G(t, s)$—that is, $\int_a^b G(t, s)f(s)\,ds = 0$, $t \in [a, b]$—this would imply that $x(t) \equiv 0$, as results from (8.46), is a solution of the nonhomogeneous equation (8.41), which is obviously a contradiction. Consequently, no function $f(t) \not\equiv 0$ on $[a, b]$ can be orthogonal to $G(t, s)$; therefore, $G(t, s)$ is complete.

The preceding considerations imply that *there exist infinitely many eigenvalues of the Sturm–Liouville problem—say,* $\{\lambda_k\}$, $k = 1, 2, \ldots,$ *—that they are real, and that the orthonormal system of eigenfunctions* $\{\varphi_k(t)\}$, $k = 1, 2, \ldots,$ *is complete.*

From the Hilbert–Schmidt expansion theorem, the following result can be derived.

Let $g(t)$ be a function defined on $[a, b]$, continuously differentiable of the second order and satisfying the boundary conditions (8.34). *Then*

$$g(t) = \sum_{n=1}^{\infty} (g, \varphi_n)\varphi_n(t), \qquad t \in [a, b] \tag{8.50}$$

the series being uniformly and absolutely convergent on $[a, b]$.

Indeed, if we denote $f(t) = -L(g(t))$, it follows from our assumptions that $f(t)$ is continuous on $[a, b]$, and

$$g(t) = \int_a^b G(t, s)f(s)\, ds \tag{8.51}$$

because $g(t)$ satisfies the conditions in (8.34). From (8.51), we see that $g(t)$ satisfies the condition required by the Hilbert–Schmidt expansion theorem, which proves the preceding statement.

§8.7. Reduction to an Integral Equation: The Second Case

This case corresponds to the circumstance when $\lambda = 0$ is an eigenvalue of the Sturm–Liouville problem (8.33), (8.34). To reduce our problem to an integral equation, we shall use another differential operator. Namely,

$$L_0(x) = \frac{d}{dt}\left[p(t)\frac{dx}{dt}\right] - q(t)x + \lambda_0 x \tag{8.52}$$

where λ_0 is a real number that is not an eigenvalue of our problem. That such numbers exist follows from the next statement.

Any orthonormal system of continuous functions on $[a, b]$ *is at most countable.*

Let $\{\varphi_\alpha(t)\}$ be an orthonormal system, with $\alpha \in A$, where A is an arbitrary set of subscripts. For fixed α and $\varepsilon > 0$, we can find a polynomial with rational coefficients—say, $p_\alpha(t)$—such that

$$\|\varphi_\alpha(t) - p_\alpha(t)\| < \varepsilon. \tag{8.53}$$

Inasmuch as for $\alpha \neq \beta$,

$$\|\varphi_\alpha - \varphi_\beta\|^2 = \|\varphi_\alpha\|^2 + \|\varphi_\beta\|^2 - 2(\varphi_\alpha, \varphi_\beta) = 2$$

we get

$$\|\varphi_\alpha - \varphi_\beta\| = \sqrt{2} \leq \|\varphi_\alpha - p_\alpha\| + \|p_\alpha - \varphi_\beta\| < \varepsilon + \|p_\alpha - \varphi_\beta\|$$

which implies

$$\|\varphi_\beta - p_\alpha\| > \sqrt{2} - \varepsilon > \varepsilon$$

if ε is sufficiently small. Taking into account $\|\varphi_\beta - p_\beta\| < \varepsilon$, we find that $p_\alpha \neq p_\beta$, for $\alpha \neq \beta$.

Consequently, we can establish a one-to-one mapping from our system $\{\varphi_\alpha(t)\}$, $\alpha \in A$, to a set $\{p_\alpha(t)\}$ of polynomials with rational coefficients. Because the set of all polynomials with rational coefficients is countable, it follows that our system $\{\varphi_\alpha(t)\}$ is at most countable. On the other hand, the set of eigenvalues of a Sturm–Liouville problem has the same power as the set of its eigenfunctions (which form an orthonormal system, as shown in §8.5).

The set of eigenvalues of a Sturm–Liouville problem being at most countable, it follows that there exists a real number λ_0 that is not an eigenvalue. The differential operator $L_0(x)$ will now play the same role as $L(x)$ in the preceding section.

We find that our problem is equivalent to an integral equation of the form

$$x(t) = (\lambda - \lambda_0) \int_a^b G_0(t, s)x(s)\, ds \tag{8.54}$$

inasmuch as Eq. (8.33) can be written as $L_0(x) = -(\lambda - \lambda_0)x$. If we denote by $\{\mu_k\}$, $k = 1, 2, \ldots$, the eigenvalues of the kernel $G_0(t, s)$, it follows that the eigenvalues of the Sturm–Liouville problem are $\{\mu_k + \lambda_0\}$, $k = 1, 2, \ldots$. The same conclusions we stated in §8.6 hold in the case under investigation.

PROBLEMS

1. Consider the integral equation

$$x(t) = \lambda \int_a^b k(t, s)p(s)x(s)\, ds$$

where $k(t, s)$ is a kernel of Hermitian type and $p(s)$ is a continuous positive function on $[a, b]$. Show that this equation can be reduced to an equation with Hermitian kernel by multiplying it with a suitable function depending on t. Derive the corresponding properties of eigenvalues and eigenfunctions.

2. Let us consider the following generalized Sturm–Liouville system:

$$\frac{d}{dt}\left[p(t)\frac{dx}{dt}\right] + [\lambda r(t) - q(t)]x = 0 \tag{E}$$

$$\alpha_1 x(a) + \alpha_2 \dot{x}(a) = 0, \qquad \beta_1 x(b) + \beta_2 \dot{x}(b) = 0 \tag{L}$$

where $p(t)$, $q(t)$ and the coefficients in (L) satisfy the conditions specified in §8.5 and where $r(t)$ is a continuous nonnegative function on $[a, b]$. Investigate this generalized problem following the procedure used in this chapter. In particular, prove the following property: If $x(t)$ and $y(t)$ are eigenfunctions corresponding to distinct eigenvalues, then

$$\int_b^b x(t)y(t)r(t)\,dt = 0.$$

3. Consider a Sturm–Liouville system under the conditions specified in §8.5. Prove that $\lambda = 0$ can be an eigenvalue.
Hint: Investigate the equation $\ddot{x} + \lambda x = 0$ with conditions $\dot{x}(a) = \dot{x}(b) = 0$.

4. Consider the second-order nonlinear equation $\ddot{x} = f(t, x)$ and the boundary conditions $x(a) = A$, $x(b) = B$. Show first that the boundary value problem is equivalent to another, of the same kind, with homogeneous boundary conditions (i.e., $A = B = 0$). Find an integral equation of Hammerstein type (see problem 9, Chapter 7) that is equivalent to the boundary value problem stated previously. If $f(t, x)$ is continuous and bounded for $t \in [a, b]$, $-\infty < x < \infty$, then there exists at least one solution.
Hint: If $G(t, s)$ is the Green's function of the operator $\mathscr{L}(x) = -\ddot{x}$, with boundary conditions $x(a) = x(b) = 0$, then the integral equation we are looking for can be written

$$x(t) = \int_a^b G(t, s)f(s, x(s))\,ds.$$

5. Consider the Poisson kernel

$$k(t, s) = \frac{1}{2\pi}\frac{1 - h^2}{1 - 2h\cos(t - s) + h^2}$$

where $t, s \in [-\pi, \pi]$ and $|h| < 1$. Prove that the eigenvalues of this kernel are the numbers

$$1, h^{-1}, h^{-1}, \ldots, h^{-n}, h^{-n}, \ldots$$

and the eigenfunctions are

$$1, \cos t, \sin t, \ldots, \cos nt, \sin nt, \ldots.$$

Hint: One can proceed to a direct verification. It is convenient to notice that

$$2\pi k(t, s) = 1 + 2\sum_{n=1}^{\infty} h^n \cos n(t - s)$$

Notice also the fact that the trigonometric system is complete.

6. Apply the Hilbert–Schmidt expansion theorem to solve the Fredholm equation of the first kind,

$$f(t) = \int_a^b k(t, s)x(s)\,ds.$$

Assume that $k(t, s)$ is a Hermitian kernel.

7. Let $k(t, s)$ be a Hermitian kernel. We shall say that $k(t, s)$ is a *positive* kernel if $(Kf, f) > 0$ for any $f \neq 0$. Prove that $k(t, t) \geq 0$, $t \in [a, b]$. The eigenvalues of $k(t, s)$ are all positive. Denote by $\{\varphi_n(t)\}$ the system of eigenfunctions of the positive kernel $k(t, s)$. Show that

$$k(t, s) = \sum_{n=1}^{\infty} \lambda_n^{-1}\varphi_n(t)\overline{\varphi_n(s)}$$

where the series is absolutely and uniformly convergent on $[a, b] \times [a, b]$. This result is due to Mercer.
Hint: Consider the auxiliary kernels

$$k_m(t, s) = k(t, s) - \sum_{n=1}^{m} \lambda_n^{-1} \varphi_n(t) \overline{\varphi_n(s)}$$

and prove that they are positive too. Derive then the following partial result:

$$k(t, t) = \sum_{n=1}^{\infty} \lambda_n^{-1} |\varphi_n(t)|^2.$$

Use Dini's theorem on uniform convergence of monotone sequences of continuous functions. From this partial result concerning the convergence, derive the general result stated previously.

9

Miscellanea (Nice Things)

Inasmuch as we omitted many interesting subjects and topics from our exposition, we kindly invite the reader to get acquainted with some results that complete certain aspects of the problems discussed in the preceding chapters. Of course, we make no attempt to give a full treatment or description of the subjects we shall approach. However, we hope that the topics included in this chapter will stimulate the reader's interest in further study of differential and integral equations.

§9.1. Maximum and Minimum Solutions of Scalar Differential Equations

Consider the scalar differential equation

$$\dot{x} = \omega(t, x) \tag{9.1}$$

where $\omega(t, x)$ is a continuous function on

$$D: \qquad t_0 \leq t \leq t_0 + a, \qquad |x - x_0| \leq b.$$

From Peano's existence theorem, we know that there exists at least one solution $x = x(t)$ of (9.1), such that $x(t_0) = x_0$. This solution is defined on $t_0 \leq t \leq t_0 + \delta$, with $\delta = \min\{a, bM^{-1}\}$, where $M = \sup|\omega(t, x)|$, $(t, x) \in D$.

We now want to prove the following: *There exist a maximum solution*

$x_M(t)$ and a minimum solution $x_m(t)$, both passing through (t_0, x_0) and defined on $[t_0, t_0 + \tilde{\delta}]$, $\tilde{\delta} > 0$, such that

$$x_m(t) \le x(t) \le x_M(t), \qquad t \in [t_0, t_0 + \tilde{\delta}] \tag{9.2}$$

for any solution $x(t)$ of (9.1), with $x(t_0) = x_0$.

Let $\{\omega_n(t, x)\}$ be a sequence of continuous functions on D, such that

$$|\omega_n(t, x) - \omega_n(t, y)| \le L_n |x - y| \tag{9.3}$$

and

$$\lim_{n \to \infty} \omega_n(t, x) = \omega(t, x) \tag{9.4}$$

uniformly on D. The existence of such a sequence follows from the approximation theorem of Weierstrass.[1] In particular, $\omega_n(t, x)$ could be polynomials. Without loss of generality, we can assume that $\omega_n(t, x) \ge \omega(t, x)$, $(t, x) \in D$, $n = 1, 2, \ldots$. Indeed, if $\{\omega_n(t, x)\}$ satisfies (9.3) and (9.4), then $\{\omega_n(t, x) + \varepsilon_n\}$, with $\varepsilon_n = \sup |\omega(t, x) - \omega_n(t, x)|$, $(t, x) \in D$, satisfies also the last requirement.

Consider now the equations

$$\dot{x} = \omega_n(t, x), \qquad n = 1, 2, \ldots. \tag{9.5}$$

Each equation (9.5) has a unique solution $x = x_n(t)$, $x_n(t_0) = x_0$, where $x_n(t)$ is defined on $[t_0, t_0 + \tilde{\delta}]$, $\tilde{\delta} = \min\{a, b\tilde{M}^{-1}\}$, $\tilde{M}$ being such that $|\omega_n(t, x)| \le \tilde{M}$ on D, $n = 1, 2, \ldots$. Actually, $\tilde{M}$ can be chosen arbitrarily close to M.

It is easy to prove that the sequence $\{x_n(t)\}$ is uniformly bounded and equicontinuous on $[t_0, t_0 + \tilde{\delta}]$. We have $|x_n(t) - x_0| \le b$ and

$$|x_n(t) - x_n(s)| \le \tilde{M}|t - s|.$$

Let us denote

$$x_M(t) = \lim x_{n_k}(t) \quad \text{as} \quad k \to \infty \tag{9.6}$$

where $\{x_{n_k}(t)\}$ is a uniformly convergent subsequence of $\{x_n(t)\}$. If $x(t)$ is a solution of (9.1), defined on $[t_0, t_0 + \tilde{\delta}]$ and such that $x(t_0) = x_0$, then

$$\dot{x}(t) = \omega(t, x(t)) \le \omega_n(t, x(t))$$

from which we obtain (see §3.5)

$$x(t) \le x_n(t), \qquad t \in [t_0, t_0 + \tilde{\delta}], \qquad n = 1, 2, \ldots. \tag{9.7}$$

From (9.6) and (9.7), it follows that $x(t) \le x_M(t)$. Consequently,

[1] For instance, C. Corduneanu, *Almost Periodic Functions*, John Wiley & Sons, Inc., New York, 1968.

$x_M(t)$ as defined is the maximum solution of (9.1) through (t_0, x_0). That the maximum solution through a given point is unique can be seen as follows: If $x_M(t)$ and $\tilde{x}_M(t)$ were both maximum solutions, then we should have $x_M(t) \leq \tilde{x}_M(t)$ and $\tilde{x}_M(t) \leq x_M(t)$. Hence, $x_M(t) = \tilde{x}_M(t)$, which proves the uniqueness. A similar argument can be used to prove the existence of the minimum solution $x_m(t)$ through (t_0, x_0).

Remark. It is interesting to point out the following feature. If (t_1, x_1) is such that $t_0 < t_1 < t_0 + \tilde{\delta}$, $x_1 = x_M(t_1)$, then the maximum solution through (t_1, x_1) coincides with $x_M(t)$ for $t \geq t_1$. Indeed, if the maximum solution through (t_1, x_1)—say, $X_M(t)$—is such that $X_M(t) > x_M(t)$ for some $t > t_1$, then the solution defined by

$$x(t) = \begin{cases} x_M(t), & t \in [t_0, t_1] \\ X_M(t), & t > t_1 \end{cases}$$

passes through (t_0, x_0) and does not satisfy $x(t) \leq x_M(t)$. The contradiction shows that $x_M(t)$ is the maximum solution through any point situated on its graph.

Assume that (9.1) is given in a domain $\Omega \subset R^2$, with $\omega(t, x)$ continuous for $(t, x) \in \Omega$. Then through any point $(t_0, x_0) \in \Omega$ there passes a *unique* saturated maximum solution (at the right of t_0). The proof of the last statement can be carried out by the procedure we gave in §3.3, with the specification that any time we have to construct a continuation we must take the maximum solution.

There are several applications of the concept of maximum (or minimum) solution. Let us mention one of them here; it is related to the theory of differential inequalities. If $\omega(t, x)$ is continuous in Ω and $y(t)$ is a differentiable function such that $(t, y(t)) \in \Omega$, $t \in [t_0, t_1)$, then

$$\dot{y}(t) \leq \omega(t, y(t)) \tag{9.8}$$

implies

$$y(t) \leq x_M(t) \tag{9.9}$$

on the common interval of definition, where $x_M(t)$ denotes the maximum solution of (9.1) through $(t_0, y(t_0))$.

§9.2. A Theorem of Massera

Consider again Eq. (9.1)

$$\dot{x} = \omega(t, x)$$

where $\omega(t, x)$ continuous for $t \geq 0$, $|x| \leq M$, satisfies a condition that ensures the uniqueness through any point of the domain of definition (e.g., the local Lipschitz condition), and $\omega(t+T, x) = \omega(t, x)$ for a certain $T > 0$ [i.e., $\omega(t, x)$ is periodic in t].

Massera Theorem

If there exists a solution $x = x_0(t)$ of (9.1), defined on the half-axis $t \geq 0$ (hence, bounded), then (9.1) has at least one periodic solution.

Proof. It is obvious that $x_n(t) = x_0(t + nT)$, $n = 1, 2, \ldots$ are solutions of (9.1), defined on $t \geq 0$ and such that

$$|x_n(t)| \leq M, \qquad t \geq 0, \qquad n = 1, 2, \ldots. \tag{9.10}$$

Indeed,

$$\begin{aligned}\frac{d}{dt} x_n(t) &= \dot{x}_0(t + nT) = \omega(t + nT, x_0(t + nT)) \\ &= \omega(t, x_0(t + nT)) = \omega(t, x_n(t))\end{aligned}$$

which shows that $x_n(t)$ is a solution of (9.1). Estimate (9.10) follows from $|x_0(t)| \leq M$, $t \geq 0$. If for some n, $x_n(0) = x_{n+1}(0)$, the uniqueness property gives $x_n(t) = x_{n+1}(t)$—that is,

$$x_0(t + nT) \equiv x_0(t + \overline{n+1}\,T) \equiv x_0(t + nT + T).$$

This shows that $x_n(t)$ is a periodic solution of (9.1), with period T.

Assume now, for definiteness, that $x_0(0) < x_1(0)$. This statement implies $x_0(t) < x_1(t)$, $t > 0$, if we take into account the uniqueness. For $t = nT$ we get $x_n(0) < x_{n+1}(0)$, and, therefore, $x_n(t) < x_{n+1}(t)$, $t \geq 0$. Hence, $\{x_n(t)\}$ is an increasing sequence. From (9.10) it follows the existence of the limit

$$\lim_{n \to \infty} x_n(t) = x(t), \qquad t \geq 0. \tag{9.11}$$

On the other hand, there exists $N > 0$ such that $|\omega(t, x)| \leq N$ for $t \geq 0$, $|x| \leq M$. From (9.1) we obtain

$$|x_n(t) - x_n(s)| \leq N|t - s|, \qquad t, s \geq 0, \qquad n = 1, 2, \ldots. \tag{9.12}$$

It follows from (9.10) and (9.12) that $\{x_n(t)\}$ is uniformly bounded and equicontinuous on the half-axis $t = 0$. Consequently, we can apply Ascoli–Arzelà lemma on any bounded interval of the positive half-axis. Inasmuch as any monotone sequence containing a convergent subsequence is itself convergent, we conclude that (9.10) holds uniformly on any bounded interval. Hence, $x(t)$ is continuous on $t \geq 0$ and, moreover, is a solution of (9.1). The last statement follows easily from

$$x_n(t) = x_n(0) + \int_0^t \omega(s, x_n(s))\, ds, \qquad t \geq 0, \qquad n = 1, 2, \ldots$$

considering that $x_n(0) \to \xi$, $|\xi| \leq M$, as $n \to \infty$.

Inasmuch as $x(T) = \lim x_n(T) = \lim x_{n+1}(0)$ as $n \to \infty$, we obtain $x(T) = \xi = x(0)$. But $x(t + T)$ is also a solution of (9.1), which means that $x(t + T) \equiv x(t)$, $t \geq 0$. The theorem of Massera is thus proved.

The significance of the preceding result consists mainly in the fact that the existence of a *periodic* solution is reduced to the existence of a *bounded* solution. Of course, we have in mind periodic equations—that is, such that $\omega(t + T, x) \equiv \omega(t, x)$.

As an illustration of the Massera theorem, we shall consider Eq. (9.1) under the same conditions as before, and, in addition,

$$\omega(t, a) \leq 0, \qquad \omega(t, b) \geq 0, \qquad t \geq 0 \tag{9.13}$$

where $a < b$ are real numbers. Equations (9.13) can be written as $\dot{a} \geq \omega(t, a)$, $\dot{b} \leq \omega(t, b)$, $t \geq 0$, because $\dot{a} = \dot{b} = 0$. If x_0 is such that $a < x_0 < b$, then the solution $x(t)$ of (9.1) with $x(0) = x_0$ is defined on $t \geq 0$ and $a \leq x(t) \leq b$, $t \geq 0$. This follows easily from the main result concerning the differential inequalities.

Consequently, there exists at least one bounded solution of (9.1), defined on $t \geq 0$, which implies the existence of a periodic solution. In particular, (9.13) can be satisfied by convenient $a < 0$ and $b > 0$, when $\partial\omega/\partial x$ exists and is bounded below by a positive number m. Indeed, from

$$\omega(t, x) = \left(\frac{\partial \omega}{\partial x}\right)^* x + \omega(t, 0)$$

we find $\omega(t, a) \leq ma + \omega(t, 0) < 0$ for $a < 0$ with $|a|$ sufficiently large [because $\omega(t, 0)$ is periodic, hence bounded]. Similarly, $\omega(t, b) > 0$ for sufficiently large b.

§9.3. An Oscillation Theorem

We shall be concerned in this section with the second-order differential equation

$$\frac{d}{dt}\left[p(t)\frac{dx}{dt}\right] + q(t)x = 0, \qquad t \geq 0 \tag{9.14}$$

where $p(t) > 0$ is continuously differentiable on $t \geq 0$ and $q(t)$ is continuous. A solution $x(t)$ of (9.14), which does not vanish identically on $t \geq 0$, is called *oscillatory* if there exists a sequence $\{t_n\}$, $t_n \to \infty$, such that $x(t_n) = 0$, $n = 1, 2, \dots$.

The oscillation theorem we want to establish can be stated as follows.[2] *A sufficient condition for any solution of* (9.14) *to be oscillatory is*

$$\int^{\infty} \frac{dt}{p(t)} = \int^{\infty} q(t)\, dt = +\infty. \tag{9.15}$$

Proof. First, for any solution $x(t)$ of (9.14) such that $x(t) > 0$ on (t_1, t_2),

$$y(t) = p\dot{x}/x \tag{9.16}$$

satisfies the Riccati equation

$$\dot{y} + \frac{y^2}{p} + q = 0, \qquad t \in (t_1, t_2). \tag{9.17}$$

Indeed, from (9.16) it follows that $xy = p\dot{x}$—hence, $\dot{x}y + x\dot{y} = (p\dot{x})$ and $\dot{x} = xy/p$. From (9.14), we obtain (9.17), taking into account that $x \neq 0$.

Assume now that $x(t)$ is a solution of (9.14) with $x(t) \neq 0$ for $t \geq a > 0$. Then $y(t)$ given by (9.16) will satisfy (9.17) on $t \geq a$. Consequently, for sufficiently large t, we get from (9.17)

$$y(t) + \int_a^t \frac{y^2(s)}{p(s)}\, ds = y(a) - \int_a^t q(s)\, ds < 0. \tag{9.18}$$

Denote

$$R(t) = \int_a^t \frac{y^2(s)}{p(s)}\, ds.$$

Then (9.18) gives

$$R^2(t) < R'(t)p(t), \qquad t \geq b > a. \tag{9.19}$$

Indeed, $R'(t)p(t) = y^2(t)$, and (9.18) yields $y + R < 0$—that is, $-y > R \geq 0$ and $y^2 > R^2$, which leads to (9.19).

Separation of variables in (9.19) and integration from b to $t > b$ give

$$\int_b^t \frac{ds}{p(s)} \leq \frac{1}{R(b)} - \frac{1}{R(t)} < \frac{1}{R(b)} \tag{9.20}$$

which contradicts (9.15). Note that the result stated in the preceding theorem is essentially due to Wintner and Leighton.

Let us apply the theorem to the famous Bessel equation,

$$t^2\ddot{x} + t\dot{x} + (t^2 - \alpha^2)x = 0, \qquad t > 0. \tag{9.21}$$

[2] See W. J. Coles, *Proc. Am. Math. Soc.* **2** (1968).

A simple calculation shows that Eq. (9.21) is equivalent to

$$(t\dot{x})^{\cdot} + \left(t - \frac{\alpha^2}{t}\right)x = 0, \qquad t > 0.$$

With $p(t) = t$ and $q(t) = t - \alpha^2 t^{-1}$, $t > 0$, conditions (9.15) are obviously verified. Hence, any solution of the Bessel equation (or, as it is frequently called, any *Bessel function*) vanishes infinitely many times on the positive half-axis.

Another interesting example is given by

$$\ddot{x} + [\omega^2 + \lambda f(t)]x = 0, \qquad t \geq 0$$

where $\omega > 0$, $\int_0^t f(s)\,ds$ is bounded, and λ is a small parameter.

§9.4. A Formula of Picone and Its Consequences

The formula we are going to establish regards a pair of second-order differential equations

$$\frac{d}{dt}\left[p(t)\frac{dx}{dt}\right] - q(t)x = 0 \tag{9.22}$$

$$\frac{d}{dt}\left[p_1(t)\frac{dy}{dt}\right] - q_1(t)y = 0 \tag{9.23}$$

where $p(t) > 0$ and $p_1(t) > 0$ are continuously differentiable on $[a, b]$, whereas $q(t)$ and $q_1(t)$ are merely continuous on the same interval. Assume that $x(t)$ is a solution of (9.22) and $y(t)$ is a solution of (9.23), such that $y(t) \neq 0$ on $[a, b]$. Then

$$\frac{d}{dt}\left[\frac{x}{y}(p\dot{x}y - p_1 x\dot{y})\right] = (q - q_1)x^2 + (p - p_1)\dot{x}^2 + p_1\left(\dot{x} - \frac{x}{y}\dot{y}\right)^2. \tag{9.24}$$

Indeed, the left-hand side equals

$$x\frac{d}{dt}(p\dot{x}) + p\dot{x}^2 - 2p_1 x\dot{x}\frac{\dot{y}}{y} + p_1\frac{x^2}{y^2}\dot{y}^2 - \frac{x^2}{y}\cdot\frac{d}{dt}(p_1\dot{y})$$

$$= qx^2 + p\dot{x}^2 - 2p_1 x\dot{x}\frac{\dot{y}}{y} + p_1\frac{x^2}{y^2}\dot{y}^2 - q_1x^2$$

if we consider (9.22) and (9.23).

To obtain the formula of Picone, we shall integrate both sides of (9.24) from a to b, assuming $x(a) = x(b) = 0$. We obtain

$$\int_a^b (q - q_1)x^2\,dt + \int_a^b (p - p_1)\dot{x}^2\,dt + \int_a^b p_1\left(\dot{x} - \frac{x}{y}\dot{y}\right)^2 dt = 0 \tag{9.25}$$

inasmuch as

$$\frac{x}{y}(p\dot{x}y - p_1 x\dot{y})\Big|_a^b = 0. \tag{9.26}$$

Actually, (9.26) holds under the weaker assumption $y(t) \neq 0$, $t \in (a, b)$. Indeed, it suffices to show that

$$\lim_{t\to a+} \frac{x^2(t)}{y(t)} = 0, \qquad \lim_{t\to b-} \frac{x^2(t)}{y(t)} = 0. \tag{9.27}$$

Because $x(a) = 0$, it follows that $x^2(t) = (t-a)^2\, x_1(t)$ with $x_1(t)$ bounded at $t = a$. If $y(a) = 0$, then $y(t) = (t-a)y_1(t)$, with $y_1(a) \neq 0$. Otherwise, we could write $y(t) = (t-a)^2\, y_2(t)$, which implies $\dot{y}(a) = 0$. Now, from $y(a) = \dot{y}(a) = 0$, it follows that $y(t) \equiv 0$ on $[a, b]$ [from the uniqueness property applied to (9.23)]. Hence, the first formula (9.27) is proved. A similar argument holds with respect to the second formula (9.27).

Therefore, the following result is proved: *The formula of Picone* (9.25) *holds true for any pair of solutions* $x(t)$, $y(t)$ *of Eqs.* (9.22) *and* (9.23), *respectively, such that* $x(a) = x(b) = 0$, $y(t) \neq 0$, $t \in (a, b)$.

From the preceding result, we can derive several interesting properties concerning the solutions of second-order differential equations of the form (9.22). Assume that Eqs. (9.22) and (9.23) coincide—that is, $p(t) \equiv p_1(t)$, $q(t) \equiv q_1(t)$. If $x(t)$ vanishes at the ends of the interval (a, b) and $y(t) \neq 0$ on (a, b), then Picone's formula becomes

$$\int_a^b p\left(\dot{x} - \frac{x}{y}\dot{y}\right)^2 dt = 0.$$

Inasmuch as $p > 0$, we obtain

$$0 = \dot{x} - \frac{x}{y}\dot{y} = y\left(\frac{x}{y}\right)^{\cdot}$$

which yields $y = cx$. Hence, x and y are linearly dependent solutions of (9.22). Of course, $y(t)$ vanishes also for $t = a$ and $t = b$. Consequently, the following result concerning the separation of zeros of solutions of Eq. (9.22) was proved.

Assume $p(t)$ *positive and continuously differentiable on* $[a, b]$, *with* $q(t)$ *continuous on the same interval. If* $x(t)$ *is a solution of* (9.22) *such that* $x(a) = x(b) = 0$ *and* $y(t)$ *is another solution that is linearly independent of* $x(t)$, *then* $y(t)$ *has at least one zero in* (a, b).

Another noteworthy consequence of formula (9.25), generalizing the previous result, is the so-called *Sturm's comparison theorem*. It can be stated as follows.

If $p(t)$, $p_1(t)$, $q(t)$, $q_1(t)$ satisfy the preceding conditions, and

$$p(t) \geq p_1(t), \qquad q(t) \geq q_1(t), \qquad t \in [a, b] \tag{9.28}$$

where $q(t)$ and $q_1(t)$ cannot vanish identically on any subinterval of $[a, b]$, then between two successive zeros of an arbitrary solution of (9.22) there exists at least one zero of any solution of (9.23), unless both equations coincide and the solutions are linearly dependent.

Without loss of generality, we can assume that $x(t)$ is a solution of (9.22) with $x(a) = x(b) = 0$, $x(t) \neq 0$, $t \in (a, b)$. If such is not the case, we can restrict our considerations to a subinterval $[\alpha, \beta] \subset [a, b]$, such that $x(\alpha) = x(\beta) = 0$, $x(t) \neq 0$, $t \in (\alpha, \beta)$. Now assume that a solution $y(t)$ of (9.23) does not vanish for $t \in (a, b)$. Then Picone's identity gives $q(t) \equiv q_1(t)$, $[p(t) - p_1(t)]\dot{x}(t) \equiv 0$,and $\dot{x}(t)y(t) - x(t)\dot{y}(t) = 0$, $t \in [a, b]$. We claim that $p(t) \equiv p_1(t)$. Otherwise, we should have $x(t) = 0$ on a certain subinterval of $[a, b]$, implying $q(t) \equiv 0$ (hence, $q_1(t) \equiv 0$) on the same subinterval. Because this contradicts our conditions, it follows $p(t) = p_1(t)$ on $[a, b]$. Therefore, Eqs. (9.22) and (9.23) coincide, and from

$$\dot{x}(t)y(t) - x(t)\dot{y}(t) = 0$$

we find that $x(t)$ and $y(t)$ are linearly dependent. The Sturm's comparison theorem is thereby proved.

To illustrate the preceding results, we shall consider again the Bessel equation (9.21). If we put $y = t^{1/2}x$, then x and y vanish simultaneously for $t > 0$. For y, we obtain the following equation:

$$\ddot{y} + \left[1 + \frac{1 - 4\alpha^2}{4t^2}\right] y = 0 \tag{9.29}$$

which can be compared easily with

$$\ddot{z} + (1 + \varepsilon)z = 0, \qquad \varepsilon > 0. \tag{9.30}$$

Indeed, $\varepsilon > (1 - 4\alpha^2)/4t^2$ for sufficiently large t. Hence, the zeros of any solution of (9.30) separate those of an arbitrary solution of (9.29)—that is, of an arbitrary Bessel function.

Inasmuch as the solutions of (9.30) are given by

$$z(t) = C \sin[(1 + \varepsilon)^{1/2}t + \delta]$$

we conclude that any interval $[a, a + \pi(1 + \varepsilon)^{-1/2}]$, with a sufficiently large, contains at least one zero of any Bessel function of order α. On the other hand, $-\varepsilon < (1 - 4\alpha^2)/4t^2$ for sufficiently large t, which allows a comparison of Eqs. (9.29) and (9.31):

$$\ddot{v} + (1 - \varepsilon)v = 0. \tag{9.31}$$

The preceding procedure leads to a satisfactory localization of the zeros of an arbitrary Bessel function.

§9.5. Liénard's Equation

We shall investigate in this section the second-order nonlinear equation

$$\ddot{x} + f(x)\dot{x} + g(x) = 0 \tag{9.32}$$

which is known in the mathematical literature as *Liénard's equation.* Actually, Liénard investigated a particular case of (9.32), which corresponds to $g(x) = x$, being mainly concerned with the existence of a periodic solution. His method retains its validity under more general assumptions, and we are going to present here a result obtained by Levinson and Smith.[3]

The following conditions are assumed:

1. *$f(x)$ and $g(x)$ are continuous for all x, and, moreover, $g(x)$ satisfies a Lipschitz condition*;
2. *$f(x)$ is even, $f(0) < 0$ and $g(x)$ is odd, such that $xg(x) > 0$ for $x \neq 0$*;
3. *$F(x) = \int_0^x f(u)\,du \to \infty$ as $x \to \infty$*;
4. *$F(x)$ is monotone increasing for $x \geq a > 0$, where a is the single positive zero of $F(x)$.*

Inasmuch as Eq. (9.32) is equivalent to

$$\dot{x} = y - F(x), \qquad \dot{y} = -g(x) \tag{9.33}$$

we have to consider system (9.33) under assumptions 1 through 4. This system is autonomous (see §4.5), and the following properties hold:

(a) *It satisfies a local Lipschitz condition with respect to both variables x, y.*

Indeed, $g(x)$ is assumed locally Lipschitzian, and from $F'(x) = f(x)$, it follows that $y - F(x)$ satisfies also a local Lipschitz condition. Therefore, the local existence and uniqueness hold for (9.33), through any point in the plane (x, y).

(b) *If $(x(t), y(t))$ is a solution of (9.33), so is $(-x(t), -y(t))$.*

In other words, any curve that is symmetric with respect to the origin of a trajectory is also a trajectory of (9.33). Statement (b) follows easily from the fact that both functions $y - F(x)$ and $-g(x)$ are odd.

(c) *The slope of the tangent to the trajectory passing through (x, y) is given by*

$$\frac{dy}{dx} = \frac{-g(x)}{y - F(x)}. \tag{9.34}$$

[3] See N. Levinson and O. K. Smith, *Duke Math. Journal* (1942).

Consequently, if a trajectory of (9.33) meets the y axis ($x = 0$), then its tangent is horizontal (except the origin which is a singular point). Because $g(0) = 0$, from (9.34) we see that the slope of the tangent is zero. If we denote by Γ the graph of the function $y = F(x)$, then the tangent to a trajectory that meets Γ (at the common point) is vertical.

(d) *If* (9.33) *has a unique closed trajectory* (*cycle*), *then this trajectory must be symmetric with respect to the origin.*

Indeed, from (b) it follows that to any cycle there corresponds another that is the symmetric of the first with respect to the origin. They coincide if and only if the given cycle is itself symmetric with respect to the origin.

(e) *Assume that there exists an arc of a trajectory of* (9.33), *situated in the half-plane* $x \geq 0$, *which cuts the y axis in two symmetric points with respect to the origin. Then there exists a closed trajectory of* (9.33).

Consider the symmetric arc of the arc the existence of which is stated in (e), with respect to the origin. This arc is also part of a trajectory. If we put together the arcs of trajectory under discussion, we find a whole trajectory that is obviously closed (the uniqueness property holds; see §4.5). We are now able to prove the existence of a cycle for Liénard's equation.

Theorem

If conditions 1 *through* 4 *hold, then there exists a single closed trajectory of the system* (9.33).

Proof. As discussed, it suffices to show the existence of a single arc of trajectory, situated in the half-plane $x \geq 0$ and such that it cuts the y axis in two symmetric points with respect to the origin.

Inasmuch as $\dot{y} = -g(x) < 0$ for $x > 0$, it follows that y decreases along any arc of trajectory situated to the right of the y axis. On the other hand, from $\dot{x} = y - F(x)$ we see that x increases along any part of trajectory above Γ ($y = F(x)$) and decreases otherwise.

We shall prove first that any trajectory entering the half-plane $x > 0$ can only go down meeting Γ at some point $(\alpha, F(\alpha))$, $\alpha > 0$, remaining beyond $(\alpha, F(\alpha))$ below Γ, until it reaches the y axis. Afterward, we shall prove that there exists a single arc of trajectory such that the points of entrance and egress, with respect to the half-plane $x > 0$, arc symmetric with respect to the origin.

Indeed, if the trajectory $y = y(x)$ enters the half-plane $x > 0$ through the point $(0, y_0)$, $y_0 > 0$ (for $y_0 < 0$, it can only leave the half-plane), it must meet Γ because $y = F(x) \to \infty$ as $x \to \infty$, whereas $y = y(x)$ decreases [see Eq. (9.34)]. Hence, there exists $\alpha > 0$ such that our trajectory will meet Γ

at $(\alpha, F(\alpha))$. It is convenient to denote by C_α the preceding trajectory. We know from (c) that the tangent to C_α at $(\alpha, F(\alpha))$ is vertical. The fact that beyond $(\alpha, F(\alpha))$ C_α remains below Γ can be easily seen if we consider that on any part of the trajectory above Γ x increases, whereas y is always decreasing. Furthermore, on that part of C_α situated below Γ we have $0 \leq g(x) \leq M$ and $F(x) - y \geq \delta > 0$ if $0 \leq x \leq \alpha - \varepsilon$, $\varepsilon > 0$. Consequently, dy/dx is bounded, which implies that $y = y(x)$ is bounded on $[0, \alpha - \varepsilon]$. Inasmuch as $y = y(x)$ is monotone decreasing, it follows that $\lim y(x)$ exists as $x \to 0+$. Hence, C_α again meets the y axis at a certain point $(0, y_1)$, with $y_1 < 0$ (see Fig. 3).

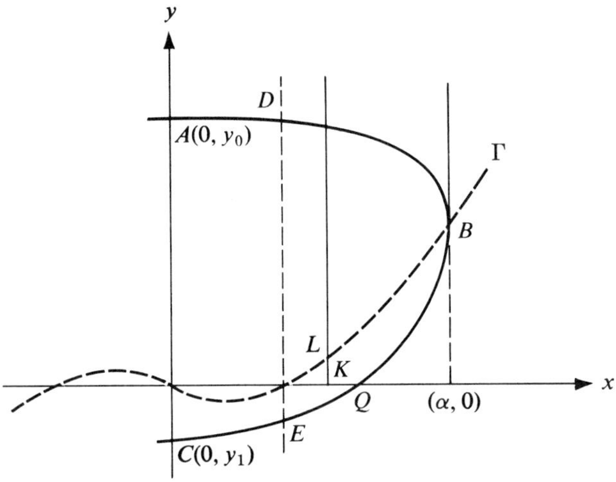

Figure 3

It remains to prove the existence of an $\alpha > 0$, such that $y_1 = -y_0$. Let us consider the function

$$u(x, y) = \tfrac{1}{2}y^2 + G(x), \qquad G(x) = \int_0^x g(u)\, du \tag{9.35}$$

and observe that $y_1 = -y_0 < 0$ takes place if and only if $u(0, y_0) = u(0, y_1)$. In other words, an arc of trajectory starting from $(0, y_0)$, situated in the half-plane $x > 0$ and ending at $(0, y_1)$, for which the variation of u [i.e., $u(0, y_1) - u(0, y_0)$] is zero, will satisfy also $y_1 = -y_0$.

Inasmuch as $du = y\, dy + g(x)\, dx$, from (9.33) we find easily

$$du = F(x)\, dy. \tag{9.36}$$

If we denote by $\varphi(\alpha)$ the variation of u along the arc of C_α situated in the half-plane $x > 0$, then (9.36) gives

$$\varphi(\alpha) = \int_{\widehat{ABC}} F(x)\, dy = u(0, y_1) - u(0, y_0). \tag{9.37}$$

A first remark is that for $\alpha \leq a$, both $F(x)$ and dy are negative—hence $\varphi(\alpha) > 0$. For such values of α, we cannot obtain the needed half-trajectory.

Assume now that $\alpha > a$. Using the notation shown in Fig. 3, let us denote

$$\varphi_1(\alpha) = \int_{\widehat{AD}} du + \int_{\widehat{EC}} du, \qquad \varphi_2(\alpha) = \int_{\widehat{DBE}} du \tag{9.38}$$

so that $\varphi(\alpha) = \varphi_1(\alpha) + \varphi_2(\alpha)$, $\alpha > a$. It is our aim to prove that both $\varphi_1(\alpha)$ and $\varphi_2(\alpha)$ are decreasing for $\alpha > a$. Indeed, along $\widehat{AD}$ or $\widehat{EC}$ we have

$$du = \frac{Fg}{F - y} dx.$$

For increasing α, arc $\widehat{AD}$ is rising and $\widehat{EC}$ is decreasing. For $\varphi_1(\alpha)$, the limits of integration are given; taking into account that $\int_{\widehat{EC}} = -\int_{\widehat{CE}}$, we conclude that $\varphi_1(\alpha)$ is decreasing with α. On arc $\widehat{DBE}$ we have $du < 0$, because $F(x) \geq 0$ and $dy < 0$. Hence, $\varphi_2(\alpha) < 0$.

To prove that $\varphi_2(\alpha)$ decreases, we shall proceed as follows. Consider the mapping $\xi = F(x)$, $\eta = y$ of the half-plane $x \geq a$ onto the half-plane $\xi \geq 0$. Because $F(x)$ is monotone and continuous for $x \geq a$, we conclude that the mapping is one-to-one, it is continuous, and so also is its inverse. Hence, $\xi = F(x)$, $\eta = y$ is a topological mapping from $x \geq a$ to $\xi \geq 0$. The arc $\widehat{DE}$ is carried by this mapping into an arc $\widehat{D_1 E_1}$ with its endpoints on the η axis, and $\varphi_2(\alpha)$ represents the area between $\widehat{D_1 E_1}$ and the η axis. The effect of increasing α is to raise D_1—say, to D_2—and to lower E_1—say, to E_2. Inasmuch as $\widehat{D_1 E_1}$ and $\widehat{D_2 E_2}$ have no common points, the result of an increase of α is an increase in the area between the arc and the η axis. But $\varphi_2(\alpha)$ equals the (oriented) area between the arc and the η axis, which shows that $\varphi_2(\alpha)$ decreases when α increases. Consequently, $\varphi(\alpha) = \varphi_1(\alpha) + \varphi_2(\alpha)$ is decreasing.

We want to show that $\varphi_2(\alpha) \to -\infty$ as $\alpha \to \infty$. Because $du < 0$ on $\widehat{DBE}$, it suffices to show that $\int du \to -\infty$ along some subarc of $\widehat{DBE}$. From Fig. 3 we see that

$$\int_{\widehat{PBQ}} du = \int_{\widehat{PBQ}} F(x)\, dy < -|PK| \times |KL| \tag{9.39}$$

if we take into account that $F(x)$ is increasing [hence, $F(x) \geq |KL|$ on $\widehat{PBQ}$ and $|\int_{\widehat{PBQ}} dy| \geq |PK|$]. Keeping K fixed in (9.39) and making $\alpha \to \infty$, we find $\varphi_2(\alpha) \to -\infty$ because $|PK|$ becomes arbitrarily large.

In summary, $\varphi(\alpha)$ is a continuous function on $\alpha > 0$, such that $\varphi(\alpha) > 0$ for $\alpha \leq a$. It is decreasing for $\alpha > a$ and $\varphi(\alpha) \to -\infty$ as $\alpha \to \infty$. Consequently, there is one and only one value of α—say, α_0, $\alpha_0 > a$—such that $\varphi(\alpha_0) = 0$. For this value of α, we have $u(0, y_1) - u(0, y_0) = 0$, which ends the proof.

Remark. By further arguments, it can be shown that the cycle the existence of which was proved in the preceding theorem is "*orbitally stable.*" The meaning of this term is intuitively clear.

As an illustration of the preceding considerations, we can mention the equation

$$\ddot{x} + \mu(x^2 - 1)\dot{x} + x = 0, \qquad \mu > 0 \tag{9.40}$$

usually known as Van der Pol's equation. Equation (9.40) is encountered in the theory of vacuum tube circuits. The existence of a nonzero periodic solution plays an important role in the applications.

§9.6. Differential Systems Without Uniqueness

The aim of this section is to investigate the properties of the differential system

$$\dot{x} = f(t, x), \qquad x, f \in R^m \tag{9.41}$$

when $f(t, x)$ is merely continuous. Assume that $f(t, x)$ is given on $\Omega \subset R^{m+1}$, where Ω is a domain. Then, through any point $(t_0, x^0) \in \Omega$ there passes at least one solution of system (9.41). This is a consequence of Peano's existence theorem (see §2.4). The solution is defined in a "small" neighborhood of t_0—say, $|t - t_0| \leq \delta$.

The concepts of continuable solution or saturated solution have the meaning we know from §3.1. It is not difficult to build up a theory which is completely similar to that from §3.3. We are not going to repeat the arguments we have used in Chapter 3. For comparison, let us mention the result concerning the existence of a saturated solution that continues a given solution $x = x(t)$, $t \in [t_0, t_1)$, of system (9.41).

If $f(t, x)$ is continuous on Ω and $x = x(t)$, $t \in [t_0, t_1)$, is a continuable solution of (9.41), *then there exists at least one saturated solution of the same system—say, $y = y(t)$, $t \in [t_0, t_2)$—such that $t_2 > t_1$ and $x(t) = y(t)$, $t \in [t_0, t_1)$.*

The behavior of a saturated solution at the ends of its interval of existence is as described in Theorem 3.3.

Let us consider now the system (9.41) with $f(t, x)$ continuous on

$$D: \qquad t_0 \leq t \leq t_0 + a, \qquad \|x - x^0\| \leq b.$$

We claim that any solution of (9.41) through (t_0, x^0) can be continued at the right until it reaches the hyperplane $t = t_0 + \delta$, where

$$\delta = \min\left\{a, \frac{b}{M}\right\}, \qquad M = \sup_D \|f(t, x)\|. \tag{9.42}$$

Indeed, such a solution can be continued until its right end lies on the boundary of D. Hence, it can be continued for those t such that $\|x(t) - x^0\| \le b$. Inasmuch as any solution $x = x(t)$ of (9.41) passing by (t_0, x^0) satisfies

$$x(t) = x^0 + \int_{t_0}^{t} f(s, x(s))\, ds \tag{9.43}$$

we find easily that $\|x(t) - x^0\| \le b$ if $t - t_0 \le \delta$.

Now, denote by S the set consisting of those points $(t, x(t))$, $t_0 \le t \le t_0 + \delta$, where $x(t)$ is a solution of (9.41) through (t_0, x^0). In other words, S is the set filled up by the integral curves of (9.41) on $[t_0, t_0 + \delta]$, starting from (t_0, x^0). Usually S is called the *integral funnel* of the system (9.41), corresponding to (t_0, x^0). Sometimes S is called the *emission zone* of (t_0, x^0), with respect to (9.41).

The integral funnel S reduces to an integral curve if and only if the uniqueness property holds true. Denote by S_τ the intersection of the funnel S with the hyperplane $t = \tau$, $t_0 \le \tau \le t_0 + \delta$. A first result we want to establish can be stated as follows.

Theorem (Peano, Kneser)

The set S_τ is a continuum (i.e., closed and connected).

Proof. Inasmuch as any solution $x(t)$ satisfies (9.43), we get $\|x(t) - x^0\| \le b$ and $\|x(t) - x(s)\| \le M|t - s|$, $t_0 \le t, s \le t_0 + \delta$. Consequently, the set of solutions constituting the funnel S is uniformly bounded and equicontinuous on $[t_0, t_0 + \delta]$. Hence, the set of solutions through (t_0, x^0) is compact in the sense of uniform convergence on $[t_0, t_0 + \delta]$. This leads immediately to the conclusion that S is a compact set in R^{m+1} and S_τ is also compact (closed and bounded).

More involved arguments are necessary to prove that S_τ is a connected set. Assume, on the contrary, that S_τ can be represented as the union of two closed sets A and B, such that $A \cap B = \phi$. Inasmuch as both A and B are compact, it follows that the distance between them is positive: $d(A, B) = \rho > 0$. Let $(\tau, x) = P$ be an arbitrary point with $\tau > t_0$, $\|x - x^0\| \le b$. Consider the function of P

$$F(P) = d(P, B) - d(P, A). \tag{9.44}$$

It can be easily seen that

$$F(P) \geq \rho, \qquad P \in A \quad \text{and} \quad F(P) \leq -\rho, \qquad P \in B. \tag{9.45}$$

Let $y(t)$, $t_0 \leq t \leq \tau$, be a solution of (9.41), starting from (t_0, x^0) and arriving at a point in A. By $z(t)$ we shall denote another solution of (9.41), starting from (t_0, x^0) and arriving at a point in B. Then (9.45) implies

$$F(\tau, y(\tau)) \geq \rho, \qquad F(\tau, z(\tau)) \leq -\rho. \tag{9.46}$$

We shall now construct a sequence of functions $\{y_n(t)\}$, associated with $y(t)$, setting $y_n(t) = y(t_{nj})$, $t_{nj} \leq t < t_{n,j+1}$, $j = 0, 1, \ldots, n-1$, where $t_{nj} = t_0 + jn^{-1}(\tau - t_0)$. We have $\lim y_n(t) = y(t)$ as $n \to \infty$, uniformly on $[t_0, \tau]$. If we put

$$\alpha_n(t) = f(t, y(t)) - f(t, y_n(t)), \qquad t \in [t_0, \tau] \tag{9.47}$$

it follows that

$$\lim_{n \to \infty} \alpha_n(t) = 0 \tag{9.48}$$

uniformly on $[t_0, \tau]$. Indeed, (9.48) follows from the fact that $f(t, x)$ is uniformly continuous on D.

Similarly, we can construct the sequence $\{z_n(t)\}$, with $\lim z_n(t) = z(t)$ as $n \to \infty$, uniformly on $[t_0, \tau]$. If

$$\beta_n(t) = f(t, z(t)) - f(t, z_n(t)), \qquad t \in [t_0, \tau] \tag{9.49}$$

then

$$\lim_{n \to \infty} \beta_n(t) = 0 \tag{9.50}$$

uniformly on $[t_0, \tau]$. Because $F(P)$ defined by (9.44) is a continuous function, it follows from (9.46) that

$$F(\tau, y_n(\tau)) > 0, \qquad F(\tau, z_n(\tau)) < 0 \tag{9.51}$$

for $n \geq N$.

Now define the sequence $\{x_n(t; u)\}$, $n = 1, 2, \ldots$, $t \in [t_0, \tau]$, $u \in [0, 1]$, as follows:

$$\gamma_n(t; u) = (1 - u)\alpha_n(t) + u\beta_n(t), \qquad n = 1, 2, \ldots \tag{9.52}$$

$$x_n(t; u) = \begin{cases} x^0, \ t_0 \leq t < t_{n1}, \\ x^0 + \int_{t_0}^{t_{n1}} [f(s, x_n(s; u)) + \gamma_n(s; u)]\, ds, \ t_{n1} \leq t < t_{n2}, \\ \cdots\cdots\cdots\cdots\cdots\cdots \\ x^0 + \int_{t_0}^{t_{nk}} [f(s, x_n(s; u)) + \gamma_n(s; u)]\, ds, \ t_{nk} \leq t < t_{n,k+1} \\ \cdots\cdots\cdots\cdots\cdots\cdots \end{cases} \tag{9.53}$$

Inasmuch as $\gamma_n(t; u)$ are piecewise continuous in t and continuous in u, such that $\lim \gamma_n(t; u) = 0$ as $n \to \infty$, uniformly on $[t_0, \tau] \times [0, 1]$, it follows from (9.53) that $x_n(t; u)$ are piecewise continuous in t and continuous in u, $u \in [0, 1]$. We have, obviously, $x_n(t; 0) = y_n(t)$, $x_n(t; 1) = z_n(t)$, $n = 1, 2, \ldots$.

Consider now the function of u, $F_{n,\tau}(u) = F(u, x_n(\tau; u))$, $u \in [0, 1]$. It is a continuous function, and (9.51) implies $F_{n,\tau}(0) = F(\tau, y_n(\tau)) > 0$, $F_{n,\tau}(1) = F(\tau, z_n(\tau)) < 0$, $n \geq N$. Consequently, there exists u_n, $0 < u_n < 1$, such that $F_{n,\tau}(u_n) = 0$.

The sequence $\{x_n(t; u_n)\}$, $n \geq N$, consists of piecewise continuous functions on $[t_0, \tau]$. To show that $\{x_n(t; u_n)\}$ contains a uniformly convergent subsequence, converging to a continuous limit function, we shall associate to $\{x_n(t; u_n)\}$ the sequence of continuous functions on $[t_0, \tau]$

$$\tilde{x}_n(t) = x^0 + \int_{t_0}^{t} [f(s, x_n(s; u_n)) + \gamma_n(s; u_n)]\, ds, \qquad n = 1, 2, \ldots. \quad (9.54)$$

It can be verified easily that $\{\tilde{x}_n(t)\}$ is uniformly bounded and equicontinuous if we take into account the properties of f and γ_n. Consequently, there exists a uniformly convergent subsequence of $\{\tilde{x}_n(t)\}$, such that the limit function—say, $x(t)$—is continuous on $[t_0, \tau]$. But for fixed $t \in [t_0, \tau]$ and n, we can find j, $0 \leq j \leq n - 1$, such that $t_{nj} \leq t < t_{n, j+1}$. Hence,

$$\|\tilde{x}_n(t) - x_n(t; u_n)\| \leq \int_{t_{nj}}^{t} \|f + \gamma_n\|\, ds \leq (M + 1)(t - t_{nj})$$

for sufficiently large n. Inasmuch as $t - t_{nj} < n^{-1}(\tau - t_0)$, we find that $\{x_n(t; u_n)\}$ has a uniformly convergent subsequence on $[t_0, \tau]$, and the limit function $x(t)$ is continuous on $[t_0, \tau]$.

From (9.53), it follows that $x(t)$ satisfies the integral equation (9.43)—that is, is a solution of (9.41) that passes through (t_0, x^0). But $F_{n,\tau}(u_n) = 0$, $n = 1, 2, \ldots$, lead to $F(\tau, x(\tau)) = 0$. This relation contradicts (9.45) and the theorem is thereby proved.

Remark. The last statement of the theorem—that is, the connectedness of S_τ—is due to Kneser. The proof is that given by K. Hayashi.[4] Another result is due to Hukuhara.

Theorem

Let (τ, ξ) be a point on the boundary of S. Then there exists a solution of (9.41) connecting (t_0, x^0) and (τ, ξ), such that it consists only of points belonging to the boundary of S.

[4] K. Hayashi, *Memoirs of the College of Science*, University of Kyoto, Japan, 1954.

The proof leans upon the preceding theorem and needs similar arguments. We shall only sketch it. First, it can be proved by standard arguments that for any τ', such that $t_0 < \tau' < \tau$, there exists a solution of (9.41) that starts from a certain boundary point of $S_{\tau'}$, and arrives at $P = (\tau, \xi)$. For any natural n, consider the subdivision of $[t_0, \tau]$ given by the equidistant points t_{nj}, $j = 0, 1, \ldots, n$, $t_{n0} = t_0$, $t_{nn} = \tau$. Now, we can construct an approximate solution of our problem as follows: let P_{n-1} be a boundary point of S, situated on the hyperplane $t = t_{n,n-1}$, such that there exists an arc of a trajectory with endpoints at P_{n-1} and P. We consider a point P_{n-2}, playing the same role for P_{n-1} as P_{n-1} plays for P, and so on. We get a solution passing by a sequence of boundary points $P_1, P_2, \ldots, P_{n-1}$ of S and connecting $P_0 = (t_0, x^0)$ with P. Such a solution does not lie, generally, on the boundary of S. However, it can be considered as an approximate solution because it is nearby the boundary of S. The set of these solutions, corresponding to arbitrary n's, is uniformly bounded and equicontinuous on $[t_0, \tau]$. This remark leads immediately to the existence of the needed solution.

§9.7. Abel's Integral Equation

In 1826, Abel found an integral equation by starting from a problem in mechanics. He gave the solution using an elegant method that is still of significance. Actually, we shall consider a more general class of integral equations, usually known as *generalized Abel equations*. They are Volterra integral equations of the first kind but their main feature consists in the fact that the kernels are essentially discontinuous functions. More precisely, we shall be concerned with integral equations of the form

$$\int_0^t \frac{k(t, s)}{(t - s)^\alpha} x(s)\, ds = f(t), \qquad 0 < \alpha < 1 \tag{9.55}$$

where $k(t, s)$ and $f(t)$ are continuous functions for $0 \le s \le t \le T$.

Assume there exists a solution of (9.55) that is continuous on $(0, T]$ and integrable. From (9.55), we easily derive

$$\int_0^t \frac{du}{(t - u)^{1-\alpha}} \cdot \int_0^u \frac{k(u, s)}{(u - s)^\alpha} x(s)\, ds = \int_0^t \frac{f(u)\, du}{(t - u)^{1-\alpha}}.$$

The order of integration can be changed, and we find

$$\int_0^t x(s)\, ds \cdot \int_s^t \frac{k(u, s)\, du}{(t - u)^{1-\alpha}(u - s)^\alpha} = \int_0^t \frac{f(u)\, du}{(t - u)^{1-\alpha}}. \tag{9.56}$$

Let us denote

$$h(t, s) = \int_s^t \frac{k(u, s)\, du}{(t-u)^{1-\alpha}(u-s)^{\alpha}}. \tag{9.57}$$

The function $h(t, s)$ is continuous and it can be also represented by the formula

$$h(t, s) = \int_0^1 \frac{k[s + v(t-s), s]}{(1-v)^{1-\alpha} v^{\alpha}}\, dv. \tag{9.58}$$

From (9.56) and (9.57), it follows that

$$\int_0^t h(t, s) x(s)\, ds = F(t) \tag{9.59}$$

where

$$F(t) = \int_0^t \frac{f(u)\, du}{(t-u)^{1-\alpha}}. \tag{9.60}$$

Equation (9.59) is a Volterra equation of the first kind, with continuous kernel. Hence, any solution of (9.55) satisfies also (9.59). The converse is also true. If $x(t)$ is a solution of (9.59), we put

$$y(t) = \int_0^t \frac{k(t, s)}{(t-s)^{\alpha}} x(s)\, ds - f(t). \tag{9.61}$$

From (9.61) it follows, taking into account (9.56), (9.57), (9.59), and (9.60), that

$$\int_0^t \frac{y(s)}{(t-s)^{1-\alpha}}\, ds = 0 \qquad 0 \le t \le T \tag{9.62}$$

It suffices now to prove that (9.62) implies $y(t) \equiv 0$. Indeed, multiplying (9.62) by $1/(u-t)^{\alpha}$ and integrating from 0 to u, we get, after changing the order of integration,

$$\int_0^u y(s)\, ds \int_s^u \frac{dt}{(u-t)^{\alpha}(t-s)^{1-\alpha}} \equiv 0 \tag{9.63}$$

which means that $\int_0^u y(s)\, ds \equiv 0$—hence, $y(t) \equiv 0$. The integral

$$\int_s^u \frac{dt}{(u-t)^{\alpha}(t-s)^{1-\alpha}} = \int_0^1 \frac{dv}{(1-v)^{\alpha} v^{1-\alpha}} = \frac{\pi}{\sin \alpha\pi}$$

is related to Eulerian functions and can be found in several books.[5]

[5] See E. D. Rainville, *Special Functions*, The Macmillan Company, New York, 1960, p. 21.

Consequently, Eqs. (9.55) and (9.59) are equivalent. Under stronger conditions (than continuity) for $k(t, s)$, Eq. (9.59) can be solved by the method given in §6.4.

To illustrate the preceding considerations, assume that $k(t, s) \equiv 1$. This leads to the equation

$$\int_0^t \frac{x(s)\,ds}{(t-s)^\alpha} = f(t) \tag{9.64}$$

which differs from the Abel's original equation only because α appears instead of $\frac{1}{2}$. We have $h(t, s) = \pi/\sin \alpha\pi$, and (9.59) becomes

$$\int_0^t x(s)\,ds = \frac{\sin \alpha\pi}{\pi} F(t).$$

Therefore,

$$x(t) = \frac{\sin \alpha\pi}{\pi} F'(t) = \frac{\sin \alpha\pi}{\pi}\left[\frac{f(0)}{t^{1-\alpha}} + \int_0^t \frac{f'(s)\,ds}{(t-s)^{1-\alpha}}\right]$$

which represents the solution of (9.64).

References

References

A. Books on Real and Complex Analysis

1. AHLFORS, LARS: *Complex Analysis*, 2nd ed., McGraw-Hill Book Company, New York, 1966.
2. APOSTOL, T. M.: *Mathematical Analysis*, 2nd ed., Addison-Wesley Publishing Co., Inc., Reading, Mass., 1974.
3. BERS, LIPMAN: *Calculus*, Holt, Rinehart & Winston, Inc., New York, 1969.
4. CARTAN, HENRI: *Elementary Theory of Analytic Functions of One or Several Complex Variables*, Addison-Wesley Publishing Co., Inc., Reading, Mass., 1973.
5. RUDIN, WALTER: *Principles of Mathematical Analysis*, 3rd ed., McGraw-Hill Book Company, New York, 1976.

B. Books on Differential Equations (Elementary and Intermediate)

1. BELLMAN, RICHARD E.: *Modern Elementary Differential Equations*, Addison-Wesley Publishing Co., Inc., Reading, Mass., 1971.

2. Boyce, W. E., and R. C. DiPrima: *Elementary Differential Equations*, 3rd ed., John Wiley & Sons, Inc., New York, 1976.

3. Brauer, F., and J. A. Nohel: *Elementary Differential Equations*, W. A. Benjamin, Inc., New York, 1968.

4. ———: *Ordinary Differential Equations: A First Course*, 2nd ed., W. A. Benjamin, Inc., New York, 1973.

5. Coddington, E. A.: *An Introduction to Ordinary Differential Equations*, Prentice-Hall, Inc., Englewood Cliffs, N. J., 1964.

6. Cole, R. H.: *Theory of Ordinary Differential Equations*, Appleton-Century-Crofts, Inc., New York, 1968.

7. Kaplan, Wilfred: *Ordinary Differential Equations*, Addison-Wesley Publishing Co., Inc., Reading, Mass., 1958.

8. Leighton, Walter: *Introduction to the Theory of Differential Equations*, McGraw-Hill Book Company, New York, 1952.

9. Martin, W. T., and E. Reissner: *Elementary Differential Equations*, Addison-Wesley Publishing Co., Inc., Reading, Mass., 1961.

10. Petrowski, I. G.: *Vorlesungen über die Theorie der gewöhnlichen Differentialgleichungen*, B. G. Teubner, Leipzig, 1954.

11. Pontrjagin, L. S.: *Ordinary Differential Equations*, Addison-Wesley Publishing Co., Inc., Reading, Mass., 1962.

12. Rainville, E. D.: *Intermediate Differential Equations*, 2nd ed., New York, 1972.

13. Rainville, E. D., and Phillip E. Bedient: *Elementary Differential Equations*, The Macmillan Company, New York, 1974.

14. Sanchez, David A.: *Ordinary Differential Equations and Stability Theory: An Introduction*, W. H. Freeman & Co., San Francisco, 1968.

15. Spiegel, Murray R.: *Applied Differential Equations*, 2nd ed., Prentice-Hall, Inc., Englewood Cliffs, N. J., 1967.

16. Stepanow, W. W.: *Lehrbuch der Differentialgleichungen*; VEB Deutscher Verlag der Wissenschaften, Berlin, 1956.

C. Books on Differential Equations (Intermediate to Advanced)

1. Birkhoff, Garrett, and Gian-Carlo Rota: *Ordinary Differential Equations*, 2nd ed., John Wiley & Sons, Inc., New York, 1969.

2. Brauer, F., and J. A. Nohel: *The Qualitative Theory of Ordinary Differential Equations*, W. A. Benjamin, Inc., New York, 1968.

3. CESARI, LAMBERTO: *Asymptotic Behavior and Stability Problems in Ordinary Differential Equations*, 2nd ed., Springer-Verlag, New York, 1963.

4. CODDINGTON, E. A., and N. LEVINSON: *Theory of Ordinary Differential Equations*, McGraw-Hill Book Company, New York, 1955.

5. HARTMAN, PH.: *Ordinary Differential Equations*, John Wiley & Sons, Inc., New York, 1964.

6. HILLE, EINAR: *Lectures on Ordinary Differential Equations*, Addison-Wesley Publishing Co., Inc., Reading, Mass., 1969.

7. HUREWICZ, W.: *Lectures on Ordinary Differential Equations*, John Wiley & Sons, Inc., New York, 1964.

8. INCE, E. L.: *Ordinary Differential Equations*, Dover Publications, Inc., New York, 1956.

9. KAMKE, E.: *Differentialgleichungen I*, Akademische Verlagsgesellschaft, Geest & Portig K.–G., Leipzig; New York, 1971.

10. LEFSCHETZ, S.: *Differential Equations: Geometric Theory*, 2nd ed., Interscience Publishers, New York, 1963.

11. NEMYTSKY, V. V., and V. V. STEPANOV: *Qualitative Theory of Differential Equations*, Princeton University Press, Princeton, N. J., 1960.

12. SANSONE, G.: *Equazioni differenziali nel campo reale*, 2nd ed., N. Zanichelli, Bologna, 1948.

13. SANSONE, G., and R. CONTI: *Non-linear Differential Equations*, Pergamon Press, New York, 1964.

D. Books on Stability Theory of Ordinary Differential Equations

1. BARBASHIN, E. A.: *Introduction to Stability Theory* (Russian), Nauka, Moscow, 1967.

2. BELLMAN, R.: *Stability Theory of Differential Equations*, Dover Publications, Inc., New York, 1969.

3. COPPEL, W. A.: *Stability and Asymptotic Behavior of Differential Equations*, D. C. Heath & Company, Boston, 1965.

4. DEMIDOVICH, B. P.: *Lectures on Mathematical Theory of Stability* (Russian), Nauka, Moscow, 1967.

5. HAHN, WOLFGANG: *Theory and Application of Liapunov's Direct Method*, Prentice-Hall, Inc., Englewood Cliffs, N. J., 1963.

6. ———: *Stability of Motion*, Springer-Verlag, Berlin, 1967.

7. La Salle, J. P., and S. Lefschetz: *Stability by Liapunov's Direct Method*, Academic Press Inc., New York, 1961.

8. Liapunov, A. M.: *Stability of Motion*, Academic Press Inc., New York, 1966.

9. Malkin, I. G.: *Theorie der Stabilität einer Bewegung*, Oldenbourg, Munich, 1959.

10. Yoshizawa, Taro: *Stability Theory by Liapunov's Second Method*, The Mathematical Society of Japan, 1966.

E. Books on Selected Advanced Topics in Differential Equations

1. Halanay, A.: *Differential Equations: Stability, Oscillations, Time Lag*, Academic Press Inc., New York, 1966.

2. Hale, Jack K.: *Oscillations in Nonlinear Systems*, McGraw-Hill Book Company, New York, 1963.

3. Hukuhara, M., T. Kimura, and T. Matuda: *Equations différentielles ordinaires du premier ordre dans le champ complexe*, The Mathematical Society of Japan, 1961.

4. Massera, J. L., and J. J. Schaffer: *Linear Differential Equations and Function Spaces*, Academic Press Inc., New York, 1966.

5. Minorsky, N.: *Nonlinear Oscillations*, D. Van Nostrand Co., Inc., Princeton, N. J., 1962.

6. Pliss, V. A.: *Nonlocal Problems of the Theory of Oscillations*, Academic Press Inc., New York, 1966.

7. Reissig, R., G. Sansone, and R. Conti: *Qualitative Theorie nichtlinearer Differentialgleichungen*, Cremonese ed., Rome, 1963.

8. ———: *Nichtlineare Differentialgleichungen höherer Ordnung*, Cremonese ed., Rome, 1969.

9. Roseau, Maurice: *Vibrations non linéaires et théorie de la stabilité*, Springer-Verlag, Berlin, 1966.

10. Urabe, Minoru: *Nonlinear Oscillations: Analytical Theory*, Academic Press Inc., New York, 1967.

F. Books on the Theory of Control Processes

1. Aizerman, A. M., and F. R. Gantmaher: *Absolute Stability of Automatic Control Systems*, Holden-Day, San Francisco, 1964.

2. Bellman, Richard: *Introduction to the Mathematical Theory of Control Processes*, vol. I, Academic Press Inc., New York, 1971.
3. Lee, Bruce, and Lawrence Markus: *Foundations of Optimal Control Theory*, John Wiley & Sons, Inc., New York, 1967.
4. Lefschetz, S.: *Stability of Nonlinear Control Systems*, Academic Press Inc., New York, 1965.
5. Letov, A. M.: *Stability in Nonlinear Control Systems*, Princeton University Press, Princeton, N. J., 1961.
6. Lur'e, A. I.: *Einige nichtlineare Probleme aus der Theorie der automatischen Regelung*, Akademie-Verlag, Berlin, 1957.
7. Pontrjagin, L. S., *et. al.*: *The Mathematical Theory of Optimal Processes*, Interscience Publishers, Inc., New York, 1962.

G. Books on Differential and Integral Equations

1. Walter, Wolfgang: *Differential and Integral Inequalities*, Springer-Verlag, Berlin, 1970.
2. Yosida, K.: *Lectures on Differential and Integral Equations*, Interscience Publishers, New York, 1960.

H. Books on Integral Equations

1. Hamel, Georg: *Integralgleichungen*, Springer-Verlag, Berlin, 1937.
2. Krasnoselsky, M. A.: *Topological Methods in the Theory of Nonlinear Integral Equations*, Pergamon Press, New York, 1964.
3. Lalesco, T.: *Introduction à la théorie des équations intégrales*, Hermann & Fils, Paris, 1912.
4. Lovitt, W. V.: *Linear Integral Equations*, Dover Publications, Inc., New York, 1924.
5. Mikhlin, S. G.: *Integral Equations and Their Applications to Certain Problems in Mechanics, Mathematical Physics, and Technology*, 2nd ed., Pergamon Press, New York, 1964.
6. Petrovskii, I. G.: *Lectures on the Theory of Integral Equations*, Graylock Press, Rochester, N. Y., 1957.
7. Tricomi, F. G.: *Integral Equations*, Interscience Publishers, New York, 1957.

8. Volterra, V.: *Leçons sur les équations intégrales et les équations intégro-différentielles*, Gauthier-Villars, Paris, 1913.

I. Books Including the Theory of Integral Equations

1. Dieudonné, J.: *Foundations of Modern Analysis*, Academic Press Inc., New York, 1969.
2. Riesz, F., and B. Sz-Nagy: *Functional Analysis*, Frederick Ungar Publishing Co., Inc., New York, 1955.
3. Smirnov, V. I.: *A Course of Higher Mathematics*, vol. 4, Addison-Wesley Publishing Co., Reading, Mass., 1964.
4. Whittaker, E. T., and G. N. Watson: *A Course of Modern Analysis*, Cambridge University Press, 1927.

Index

Index

THEORIE DES OPERATIONS LINEAIRES

By S. BANACH

—1933-63. xii + 250 pp. 5⅜x8. 8284-0110-1.

DIFFERENTIAL EQUATIONS

By H. BATEMAN

CHAPTER HEADINGS: I. Differential Equations and their Solutions. II. Integrating Factors. III. Transformations. IV. Geometrical Applications. V. Diff. Eqs. with Particular Solutions of a Specified Type. VI. Partial Diff. Eqs. VII. Total Diff. Eqs. VIII. Partial Diff. Eqs. of the Second Order. IX. Integration in Series. X. The Solution of Linear Diff. Eqs. by Means of Definite Integrals. XI. The Mechanical Integration of Diff. Eqs.

—1917-67. xi + 306 pp. 5⅜x8. 8284-0190-X.

MEASURE AND INTEGRATION

By S. K. BERBERIAN

A highly flexible graduate level text. Part I is designed for a one-semester introductory course; the book as a whole is suitable for a full-year course. Numerous exercises.

Partial Contents: PART ONE: I. Measures. II. Measurable Functions. III. Sequences of Measurable Functions. IV. Integrable Functions. V. Convergence Theorems. VI. Product Measures. VII. Finite Signed Measures. PART TWO: VIII. Integration over Locally compact Spaces (. . . The Riesz-Markoff Representation Theorem, . . .). IX. Integration over Locally Compact Groups (Topological Groups, . . . , Haar Integral, Convolution, The Group Algebra, . . .). BIBLIOGRAPHY. INDEX.

—1965-70. xx + 312 pp. 6x9. 8284-0241-8.

L'APPROXIMATION

By S. BERNSTEIN and CH. de LA VALLÉE POUSSIN

TWO VOLUMES IN ONE:

Leçons sur les Propriétés Extrémales et la Meilleure Approximation des Fonctions Analytiques d'une Variable Réelle, *by Bernstein.*

Leçons sur l'approximation des Fonctions d'une Variable Réelle, *by Vallée Poussin.*

—1925/19-69. 363 pp. 6x9. 8284-0198-5. 2 v. in 1.

LECTURES ON THE CALCULUS OF VARIATIONS

By O. BOLZA

A standard text by a major contributor to the theory. Suitable for individual study by anyone with a good background in the Calculus and the elements of Real Variables.

—2nd (corr.) ed. 1961. 280 pp. 5⅜x8. 8284-0145-4.

VORLESUNGEN UEBER VARIATIONSRECHNUNG

By O. BOLZA

A standard text and reference work, by one of the major contributors to the theory.

—1909-63. ix + 715 pp. 5⅜x8. 8284-0160-8.

THEORIE DER KONVEXEN KOERPER

By T. BONNESEN and W. FENCHEL

"Remarkable monograph."

—*J. D. Tamarkin, Bulletin of the A. M. S.*

—1934. 171 pp. 5½x8½. 8284-0054-7.

THE CALCULUS OF FINITE DIFFERENCES

By G. BOOLE

A standard work on the subject of finite differences and difference equations by one of the seminal minds in the field of finite mathematics.

Numerous exercises with answers.

—5th ed. 1970. 341 pp. 5⅜x8. 8284-1121-2.

A TREATISE ON DIFFERENTIAL EQUATIONS

By G. BOOLE

Including the Supplementary Volume.

—5th ed. 1959. xxiv + 735 pp. 5⅜x8. 8284-0128-4.

ELECTRICAL PAPERS

By O. HEAVISIDE

Heaviside's collected works are in five volumes: The two volumes of his *Electrical Papers* and the three volumes of his *Electromagnetic Theory.*

"The [forthcoming publication of my *Electromagnetic Theory*] brought the question of a reprint of the earlier papers to a crisis. For, as the later work grows out of the earlier, it seemed an absurdity to leave the earlier work behind. [It possesses] sufficient continuity of subject-matter and treatment, and even regularity of notation, to justify its presentation in the original form . . . It might be regarded . . . as an educational work for students of theoretical electricity."—*From the Preface.*

—2nd (c.) ed. 1892-1970. 1,183 pp. 5⅜x8. 8284-0235-3.
Two vol. set.

ELECTROMAGNETIC THEORY

By O. HEAVISIDE

Third edition, with an Introduction by B. A. Behrend and with added notes on Heaviside's unpublished writings. A classic since its original publication in 1894-1912.

—3rd ed. 1970. 1,610 pp. 5⅜x8. 8284-0237-X.
Three vol. set

VORLESUNGEN UEBER DIE THEORIE DER ALGEBRAISCHEN ZAHLEN

By E. HECKE

"An elegant and comprehensive account of the modern theory of algebraic numbers."
—*Bulletin of the A. M. S.*

—2nd ed. 1970. viii + 274 pp. 5⅜x8. 8284-0046-6.

INTEGRALGLEICHUNGEN UND GLEICHUNGEN MIT UNENDLICHVIELEN UNBEKANNTEN

By E. HELLINGER and O. TOEPLITZ

"Indispensable to anybody who desires to penetrate deeply into this subject."—*Bulletin of A.M.S.*

—1928-53. 286 pp. 5⅜x8. 8284-0089-X.

Grundzüge Einer Allgemeinen Theorie der LINEAREN INTEGRALGLEICHUNGEN

By D. HILBERT

—1912-53. 306 pp. 5½x8¼. 8284-0091-1.

GEOMETRY AND THE IMAGINATION

By D. HILBERT and S. COHN-VOSSEN

Translated from the German by P. Nemenyi.

"A fascinating tour of the 20th century mathematical zoo. . . . Anyone who would like to see proof of the fact that a sphere with a hole can always be bent (no matter how small the hole), learn the theorems about Klein's bottle—a bottle with no edges, no inside, and no outside—and meet other strange creatures of modern geometry will be delighted with Hilbert and Cohn-Vossen's book."
—*Scientific American.*

"Should provided stimulus and inspiration to every student and teacher of geometry."—*Nature.*

"A mathematical classic. . . . The purpose is to make the reader *see* and *feel* the proofs. . . . readers can penetrate into higher mathematics with . . . pleasure instead of the usual laborious study."
—*American Scientist.*

"Students, particularly, would benefit very much by reading this book . . . they will experience the sensation of being taken into the friendly confidence of a great mathematician and being shown the real significance of things."—*Science Progress.*

"A person with a minimum of formal training can follow the reasoning. . . . an important [book]."
—*The Mathematics Teacher.*

—1952. 358 pp. 6x9. 8284-0087-3.

GESAMMELTE ABHANDLUNGEN (Collected Papers)

By D. HILBERT

Volume I (Number Theory) contains Hilbert's papers on Number Theory, including his long paper on Algebraic Numbers. Volume II (Algebra, Invariant Theory, Geometry) covers not only the topics indicated in the sub-title but also papers on Diophantine Equations. Volume III carries the sub-title: Analysis, Foundation of Mathematics, Physics, and Miscellaneous Papers.

—1932/35-66. 1,457 pp. 6x9. 8284-0195-0.
Three vol. set

GESAMMELTE WERKE

By C. G. J. JACOBI

The complete collected works of the renowned mathematician, edited by Karl Weierstrass (Vols. 2-7), C. W. Borchardt (Vol. 1), and A. Clebsch (Vol. 8). The second edition is a corrected reprint of Vols. 1-7 of the first edition (1881/91) and an unaltered reprint of Vol. 8.

—2nd ed. 1968. 4,032 pp. 6½x9¼.
8284-0226-4. Volumes 1-7.
8284-0227-2. Volume 8.

A TREATISE ON THE LINE COMPLEX

By C. M. JESSOP

"The best introduction to the subject."—*Virgil Snyder.*

—1903-68. xv + 364 pp. 5⅜x8. 8284-0223-X.

THE CALCULUS OF FINITE DIFFERENCES

By CHARLES JORDAN

"... destined to remain the classic treatment of the subject . . . for many years to come."—*Harry C. Carver, Founder and formerly Editor of the* ANNALS OF MATHEMATICAL STATISTICS.

—3rd ed. 1965. xxi + 655 pp. 5⅜x8. 8284-0033-4.

DIFFERENTIALGLEICHUNGEN: LOESUNGSMETHODEN UND LOESUNGEN

By E. KAMKE

Everything possible that can be of use when one has a given differential equation to solve, or when one wishes to investigate that solution thoroughly.

PART A: General Methods of Solution and the Properties of the Solutions.

PART B: Boundary and Characteristic Value Problems.

PART C: Dictionary of some 1600 Equations in Lexicographical Order, with solution, techniques for solving, and references.

"A reference work of outstanding importance which should be in every mathematical library."
—*Mathematical Gazette.*

—3rd ed. 1944. 692 pp. 6x9. 8284-0044-X.